Post-Occupancy Evaluation

Post-Occupancy Evaluation

Wolfgang F. E. Preiser
University of New Mexico

Harvey Z. Rabinowitz
University of Wisconsin-Milwaukee

Edward T. White
Florida A & M University

VNR Van Nostrand Reinhold Company
New York

Library of Congress Catalog Card Number 87-15946

ISBN 0-442-27605-2

Printed in the United States of America

Designed by Beth Tondreau

Van Nostrand Reinhold Company Inc.
115 Fifth Avenue
New York, New York 10003

Van Nostrand Reinhold Company Limited
Molly Millars Lane
Wokingham, Berkshire RG11 2PY, England

Van Nostrand Reinhold
480 La Trobe Street
Melbourne, Victoria 3000, Australia

Macmillan of Canada
Division of Canada Publishing Corporation
164 Commander Boulevard
Agincourt, Ontario M1S 3C7, Canada

16 15 14 13 12 11 10 9 8 7 6 5 4 3 2 1

Library of Congress Cataloging-in-Publication Data

Preiser, Wolfgang F. E.
Post-occupancy evaluation.

Bibliography: p.
Includes indexes.
1. Architectural design—Evaluation. 2. Buildings—Psychological aspects. I. Rabinowitz, Harvey Z. II. White, Edward T. III. Title.
NA2750.P69 1987 720'.723 87-15946
ISBN 0-442-27605-2

Contents

Foreword

The twentieth century has seen tremendous change in almost all aspects of human arts and sciences. Architecture, bridging both realms, has grown and changed at an accelerated pace since the century began. By around 1950, some architects had broken rank with their historical roots, pursuing creative excursions with scant attention paid to any precedent from the past. Unfortunately, this practice too often resulted in buildings that photographed well in the slick publications of the "trade press" but that may have lost touch with their purpose of providing an appropriate setting for working, playing, healing, or learning. In part, this book is about procedures for determining whether or not design decisions made by design professionals are delivering the performance needed by those who use the building.

After World War II there was a gradual growth in the body of knowledge that we now commonly call social or behavioral science. Some branches of this science, intended to deepen our understanding of human relationships with one another and with our environment, have made genuine progress. This book is about one of these branches: post-occupancy evaluation (POE).

How did the knowledge base and methods of POE come into being, how are they used by those who have become skilled in this process, and how can others learn to use them? The authors, who have performed many POEs of buildings, provide a guide to those who want to improve their professional services by adding the POE dimension to their practice.

My own work, as director of the Building Research Board of the National Research Council, includes serving on an active committee on methods for improving the use of POE by federal agencies only. In the past few years there has been a definite increase in the use of POE by federal agencies, but there is now a need to raise the quality of these efforts.

It may seem obvious to the observer of architecture that those who practice within the profession should visit their buildings from time to time to see how well the buildings are working and thus to improve their skills. That is why a professional athlete, actor, or opera singer uses a coach: as someone to observe how well he or she is performing and to provide the feedback that is needed to do a better job. But most architects have not made a practice of visiting their creations after completion and user occupation—at least not in any systematic way. This book provides guidance on how the systematic evaluation of buildings in use can enable the design community to take advantage of the lessons learned from successful and unsuccessful building performance. Those who learn from their own history are not as likely to repeat the mistakes of the past and are certainly in a better position to make good, professional judgments in the future.

JOHN P. EBERHARD

Preface

This book is about the assessment of building performance through post-occupancy evaluation, or POE. Post-occupancy evaluation is a phase in the building process that follows the sequence of planning, programming, design, construction, and occupancy of a building. According to Kantrowitz et al. (1986), the goals and applications of POEs can vary widely. For example, POEs are used for:

- Feedback to the evaluated building for purposes of immediate problem solving
- Troubleshooting during the shakedown period, that is, after the move-in, thereby correcting unforeseen problems in building use
- Balancing and fine-tuning of the building and its use through continuous feedback
- Auditing or other focused inquiries into select aspects of building performance, such as space utilization
- Documentation of successes and failures in building performance, thus justifying new construction or remodeling of existing buildings
- Generalization of POE information for updating and improving state-of-the-art design criteria and guideline literature for the architectural profession

The focus of this book is upon the ***performance*** *of* ***occupied*** *buildings.*

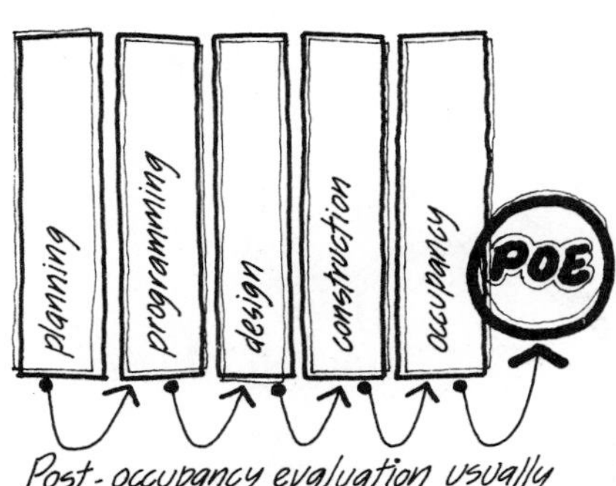

Post-occupancy evaluation usually follows ***all*** *the* ***other*** *major steps of project delivery.*

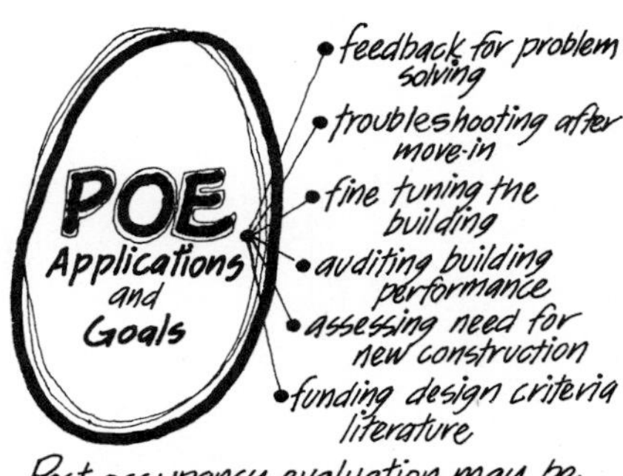

Post-occupancy evaluation may be used for ***any number*** *of* ***purposes.***

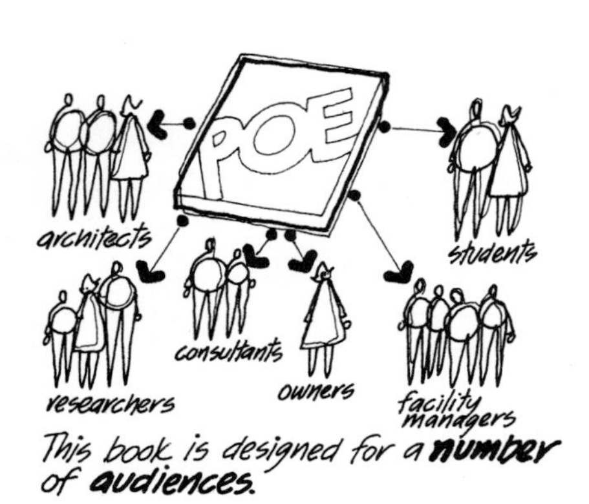

This book is designed for a **number** of **audiences**.

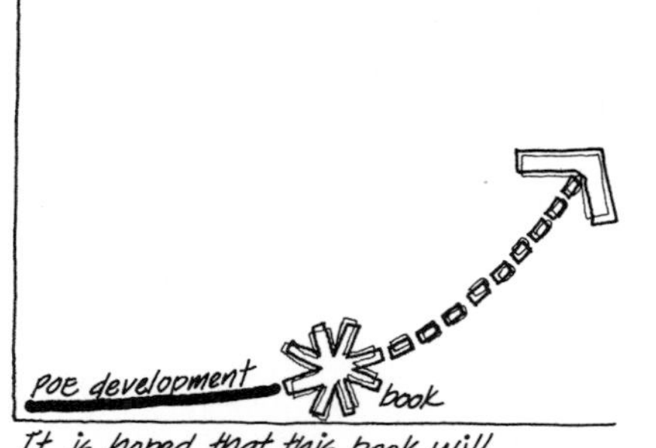

It is hoped that this book will contribute to the **continued** development of post-occupancy evaluation as a **discipline**.

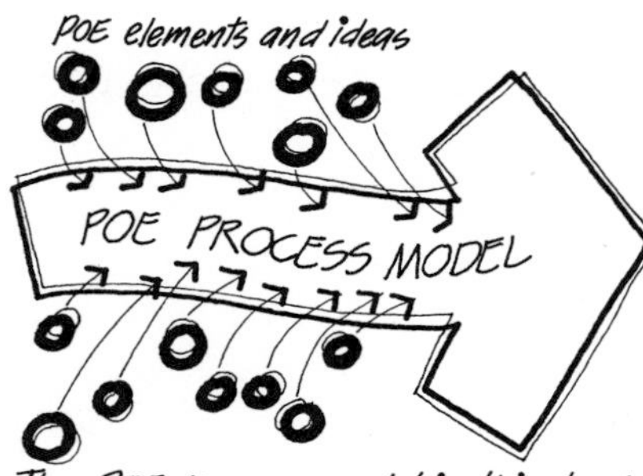

The POE process model in this book incorporates a number of elements and ideas from **others** in the field.

This book is intended to provide architects, researchers, consultants, building owners, and facilities managers with useful guidance on the process and content of POEs. It is also designed to serve as a general, introductory text in teaching this subject. *Post-Occupancy Evaluation* includes a review of the evolution of the field, a conceptual framework for POE, and pragmatic information on planning, conducting, and reporting POEs.

POE is now beginning to have a decided impact in the design and building professions, and it is hoped that others in the building industry will find the ideas and issues presented here stimulating and useful. Since POE is a relatively new field of specialization, this book attempts to serve as a catalyst for its continued development. Given that there may be several alternative approaches to POEs, the process model for POE presented in this book is comprehensive: POE elements, as evaluated by a variety of researchers in this area, have been incorporated. Different levels of effort and sophistication of POEs are proposed, ranging from single case studies to large-scale, comparative POEs on generic building types such as schools, housing, and offices.

The primary purpose of the book is to facilitate useful, economical, timely, and beneficial evaluations of buildings. This view of the subject does not permit exhaustive treatment of all methods and considerations that can be used in a POE. Detailed information on data-collection methodologies, survey research, sampling methods, or research designs, for example, can be found in the publications listed in the bibliography.

Both successes and failures in building performance are considered in POEs. The purpose of POE is to seek facts and not faults and to share the

This book is intended to **facilitate** POEs that are **cost effective** and produce **usable, effective** and **timely results**.

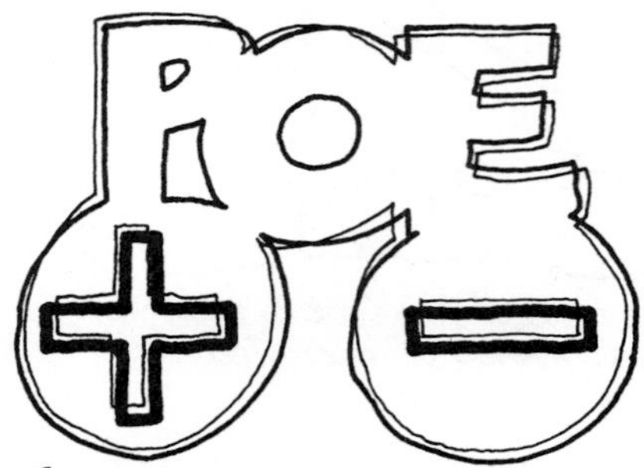

POEs should consider both the **positive** and the **negative** aspects of building performance.

findings of POEs through dissemination and the creation of information clearinghouses. The entire building industry can benefit from this information through improved building quality and better value for the dollar.

In recent years, the field of POE has expanded from the academic to the professional world. Today, POEs are conducted as part of the specialized services (Slavin 1982) being offered by a growing number of consultants, as well as a select number of progressive architectural and planning firms. Thus, POE is now becoming an established discipline that influences virtually all phases of the building process (fig. P-1).

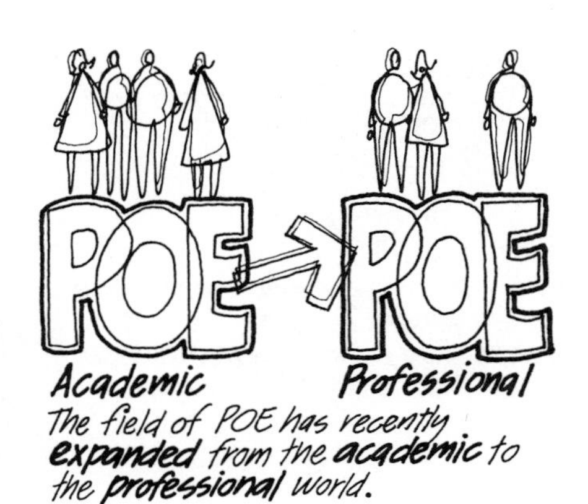

The field of POE has recently **expanded** from the **academic** to the **professional** world.

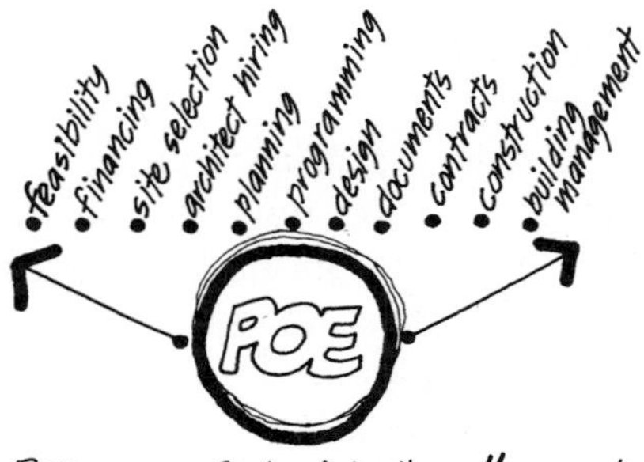

POE now affects virtually **all** aspects of the **building process.**

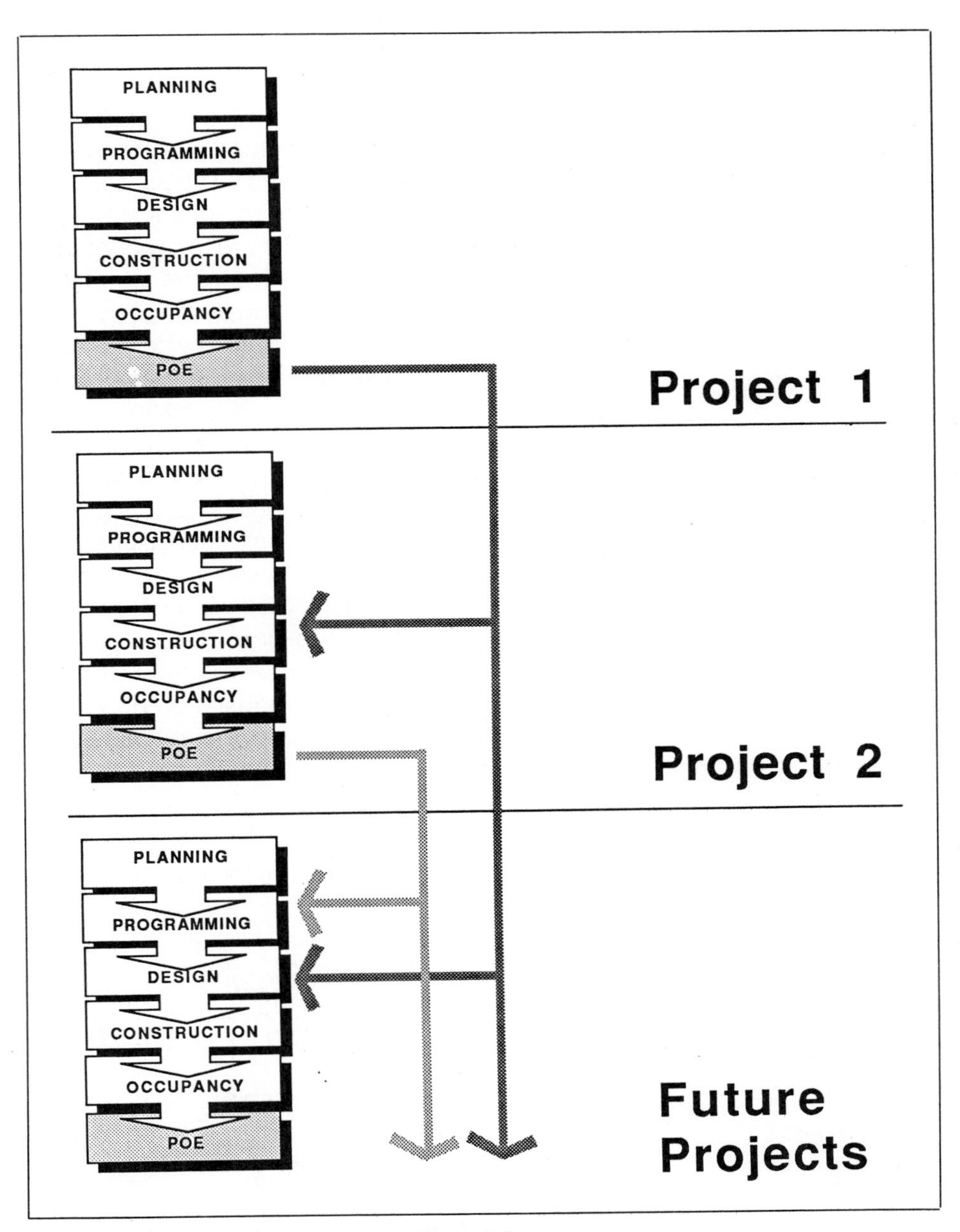

Figure P-1. POE influence on phases of the building process.

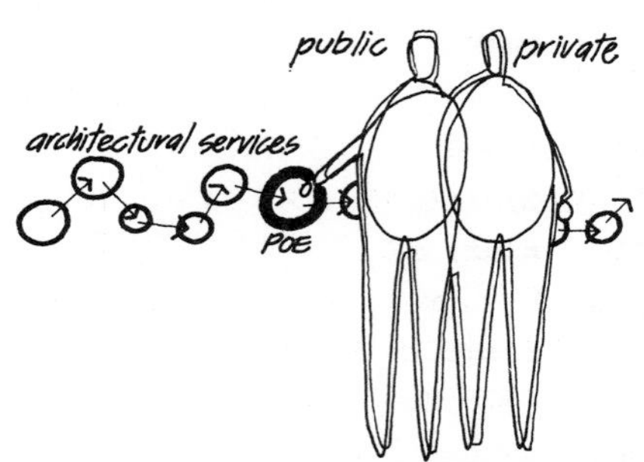

POE has now become an ***expected*** *architectural* ***service*** *by both* ***public*** *and* ***private*** *clients.*

POE is now accepted among ***building professionals*** *because it helps to ensure good building* ***performance*** *under* ***tight budget*** *conditions.*

Like architectural programming in the 1960s, POE has become part of the services required by public and private sector clients. POE is now also accepted by building professionals, partly because of the increased need for accountability and the desire to procure the best possible building performance under what are usually stringent budgetary conditions.

This is appropriate in light of the current emphasis on quality control that not only examines the performance of the facility resulting from the building process, but also the transactions that take place throughout that process, including initial feasibility, programming, design, construction, and occupancy, as well as operation and maintenance during the facility's life cycle.

Post-occupancy evaluation is the most commonly used term for the activity of evaluating buildings in use; however, there is still some controversy and confusion among professionals about the appropriate terminology. For example, the term *building diagnostics* has been introduced (Building Research Board, 1983) as an umbrella concept for comprehensive building evaluations that subsume POEs. In building diagnostics, both technical and occupant-related performance are evaluated against explicit performance criteria, with the added diagnostic-predictive dimension allowing assessment of the probable future performance of a building. The definition of POE herein excludes purely technical evaluation, for example, of heating systems or new building materials. Technical elements of building performance are considered only in terms of their effect on occupant health, safety, security, functional performance, and psychological/physical comfort.

POE assists in the ***quality control*** *of both the* ***building*** *and the* ***building process.***

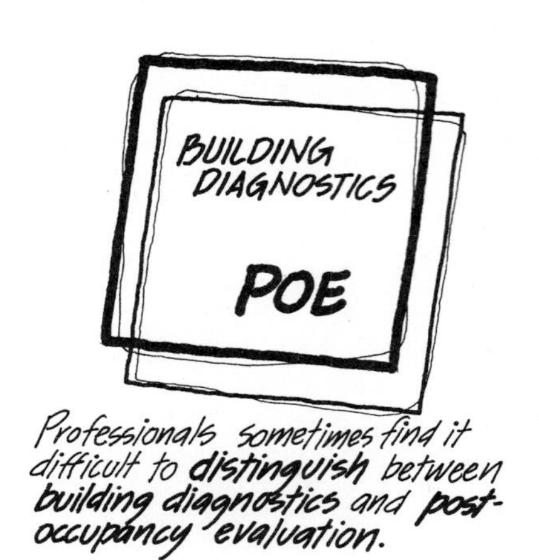

Professionals sometimes find it difficult to ***distinguish*** *between* ***building diagnostics*** *and* ***post-occupancy evaluation.***

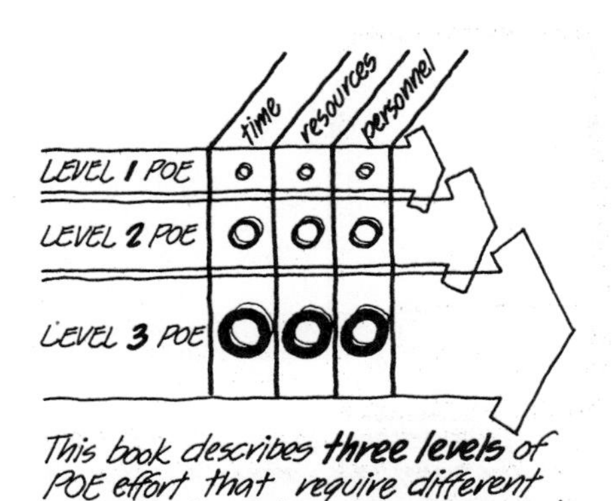

This book describes **three levels** of POE effort that require different investments of **time, resources** and **personnel**.

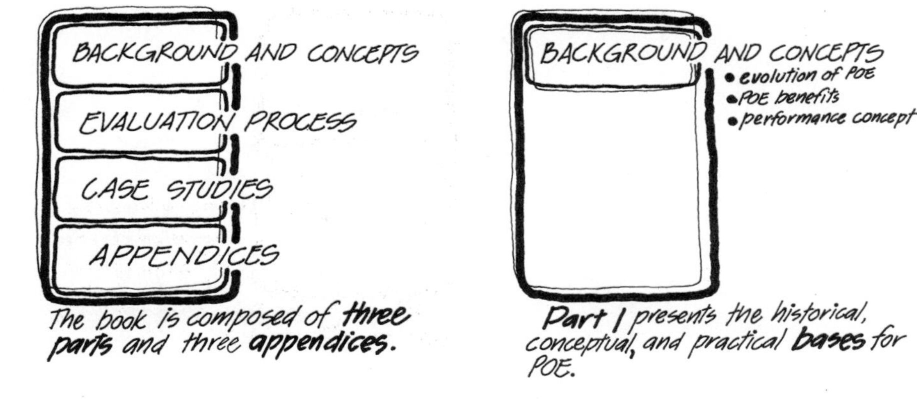

The book is composed of **three parts** and three **appendices**.

Part 1 presents the historical, conceptual, and practical **bases** for POE.

This book categorizes approaches to building evaluation by employing three levels of effort—in time, resources, and personnel required to execute POEs—complemented by case studies that illustrate practical procedures. It has three parts and three appendices. Part 1 (chapters 1 through 3), introduces the field by providing the historical, conceptual, and practical bases for POE. Chapter 1 includes a review of the evolution of POE since the mid-1960s. Chapter 2 provides the rationale for conducting POEs by examining a number of sources to determine the quality of building performance and the accountability of the design professions. The many benefits derived from POEs are described. Chapter 3 presents the *performance concept* in the building process, culminating in the *performance evaluation research framework.* In a systematic manner, this framework relates buildings and facilities to occupants and their needs. The focus is on feedback concerning the performance of existing buildings, as well as *feedforward,* or the integration of newly established design criteria, concepts, and information, into the programming and design of future, similar buildings.

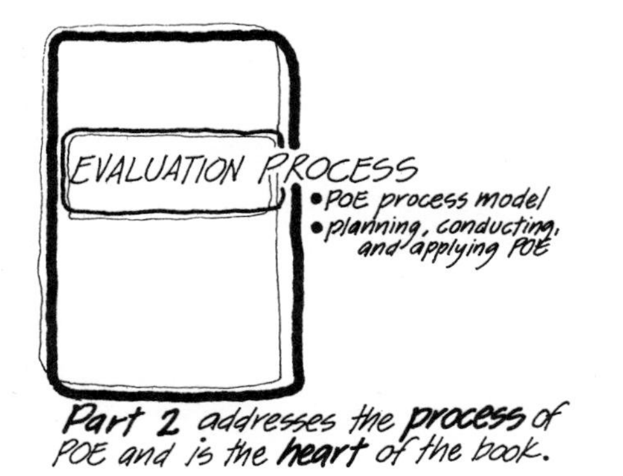

Part 2 addresses the **process** of POE and is the **heart** of the book.

Part 2 (chapters 4 through 7), dealing with the process of POE, is the core of this book. It is action-oriented and gives a detailed account of the parameters of planning, conducting, and applying POEs. Chapter 4 describes a process model for POEs. Major phases and steps in conducting POEs are outlined, based upon three basic levels of POE effort that become increasingly complex as POEs progress toward more in-depth inquiry into building performance. Chapters 5 through 7 lead the reader through the planning, conducting, and applying phases of POEs by providing guidance on the resources and logistics needed in setting up a POE, practical advice on data gathering in the field, basic information on data analysis, and the communication of findings to clients.

Part 3 (chapters 8 through 10) of the book is devoted to POE case studies at the indicative, investigative, and diagnostic levels of effort, terms that are explained in chapter 4. Chapter 8 presents indicative "walk-through"-type POEs of four senior centers in Albuquerque, New Mexico, while chapter 9 reports an investigative-type POE of an agricultural science building at the

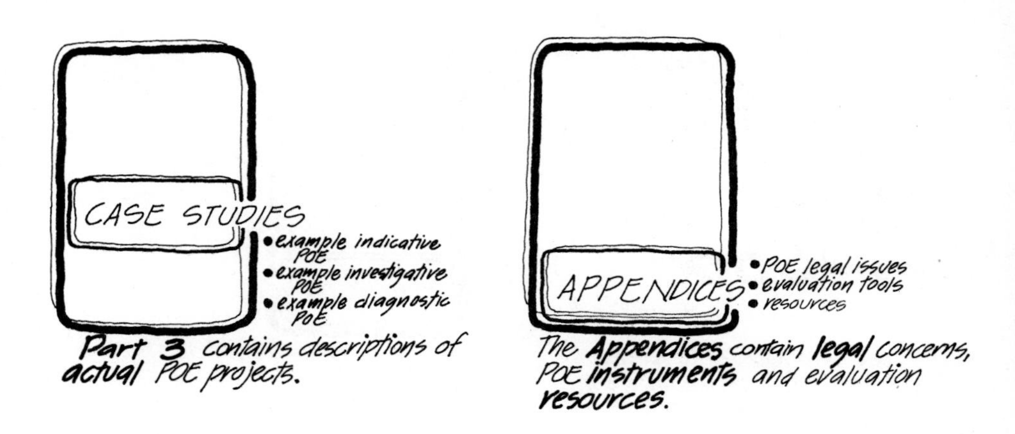

University of Kentucky. Finally, diagnostic POEs conducted on four elementary schools in Columbus, Indiana, are summarized in chapter 10.

There are three appendices; "Legal Issues in POE" (appendix A) by Robert Greenstreet; "Usable Tools" (appendix B) with evaluation instruments, checklists and reporting formats; as well as "Resources" (appendix C). The book concludes with a glossary, bibliography, and index.

Acknowledgments

The authors would like to thank all those colleagues who helped inspire research and practice in the field of post-occupancy evaluation, especially the seminal contributions of the members of the Building Performance Research Unit and their work at the University of Strathclyde, Scotland, in the 1960s and early 1970s. Their systematic approach to building evaluation marked the beginning of the field.

A number of individuals whose work and thoughts have influenced the authors' endeavors over the years includes Thomas Markus, John Daish, Heinz von Foerster, John H. Wright, Thomas Davis, John Eberhard, Michael Brill, Louis Radner, Ezra Ehrenkranz, and Roger Barker, as well as Sim van der Ryn and Murray Silverstein.

More recently, guidance and advice on this volume were provided by Robert Bechtel, Ron Goodrich, Frank Becker, Robert Marans, Steve Parshall, and Min Kantrowitz.

The ability to do innovative POEs has been most critical in advancing the thinking in this area. Support for these projects has come from the National Endowment for the Arts, the National Science Foundation, the U.S. Army Corps of Engineers, and the Irwin-Sweeny-Miller Foundation. Clients, including local school boards and municipal organizations who have provided support, as well as the hundreds of individuals whose enthusiastic cooperation and help have also been important, are gratefully acknowledged.

Thanks are owed to Robert Greenstreet for writing appendix A: "Legal Issues in POE."

The authors also thank Architectural Research Consultants, Inc., of Albuquerque, New Mexico, for assistance with those graphics in this book that were produced by laser printer. Graphics in the appendices were created by Leo L. Lucero ("Usable Tools") and Robin Benning ("Accessibility Checklist," based on a design by Thomas G. Deniston, Veterans Administration).

Finally, special thanks go to Susan Montgomery, Cecilia Fenoglio-Preiser, and Rose Mary Rabinowitz for their careful review of our manuscripts. We also thank the School of Architecture and Planning at the University of New Mexico and especially Tina Taylor for processing the words of this book.

Part 1

Post-Occupancy Evaluation: Background and Concepts

1. Uses, Benefits, and History of POE

Post-occupancy evaluation is the process of evaluating buildings in a systematic and rigorous manner after they have been built and occupied for some time. POEs focus on building occupants and their needs, and thus they provide insights into the consequences of past design decisions and the resulting building performance. This knowledge forms a sound basis for creating better buildings in the future.

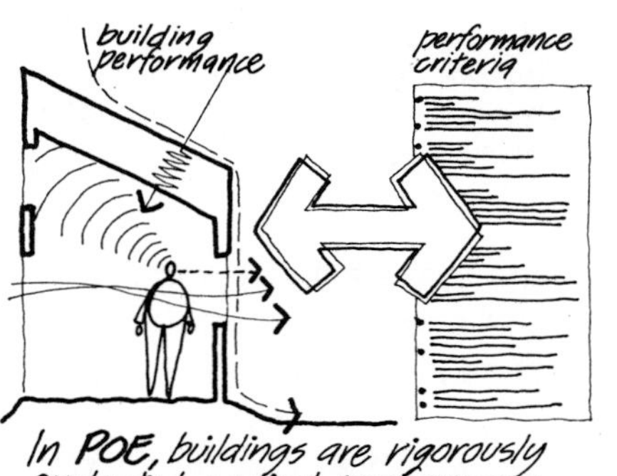

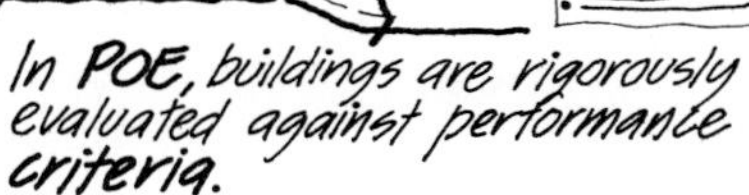
In POE, buildings are rigorously evaluated against performance **criteria**.

Normally, POEs focus on the building **occupants** and their **needs**.

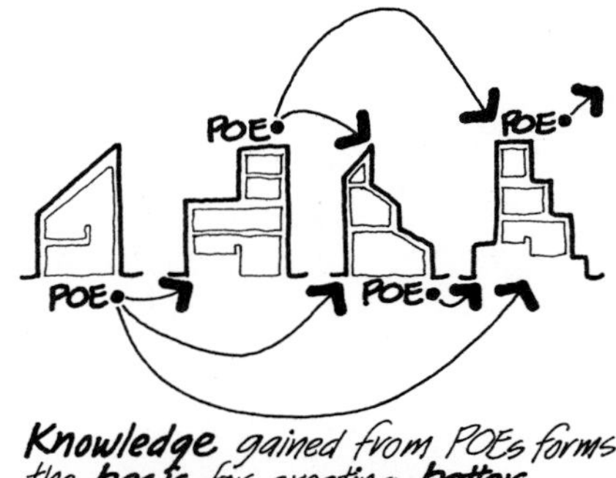

Knowledge gained from POEs forms the **basis** for creating **better** buildings in the **future**.

We **informally** evaluate aspects of building performance **every day**.

The performance of buildings is evaluated regularly, although not necessarily in a self-conscious and explicit way. In a hotel room, for example, conversations taking place next door may be overheard. In this case the acoustical performance of the building is being assessed. The room temperature, the quality of lighting, storage, finishes, and even the esthetic quality of the view from the hotel window are also informally evaluated.

Similarly, those waiting for an elevator may judge the waiting time to be excessive. The evaluation criteria used in this case come from expectations that are based on previous experiences with elevators.

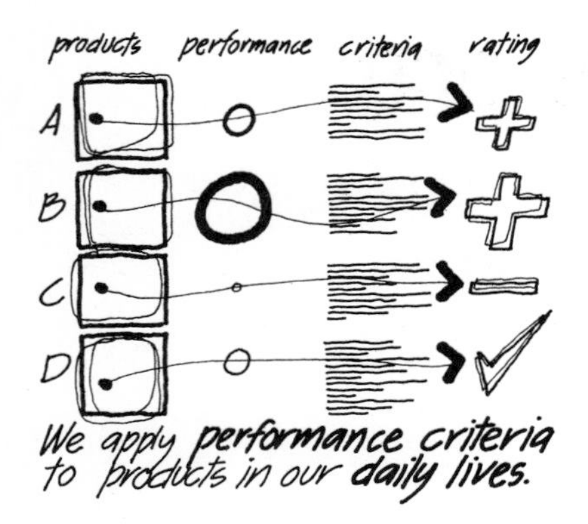

*We apply **performance criteria** to products in our **daily** lives.*

There are many products from which specific performance is expected. For instance, most people are quite careful when choosing a car. Car performance can be evaluated in terms of gas mileage, braking distance, trunk space, acceleration time, image, and so on. Along the same lines, a box camera would not be selected to produce professional-quality photographs. Instead, more sophisticated equipment is examined and evaluated according to performance criteria for shutter speeds, reliability, filters, and availability of a range of accessories.

For the buyer, a satisfactory choice is the result of an evaluation of the quality and performance of such products based on the same criteria of performance and evaluation. The goal of the manufacturer or designer is to create products with desirable features, to provide good value, and to minimize problems and failures.

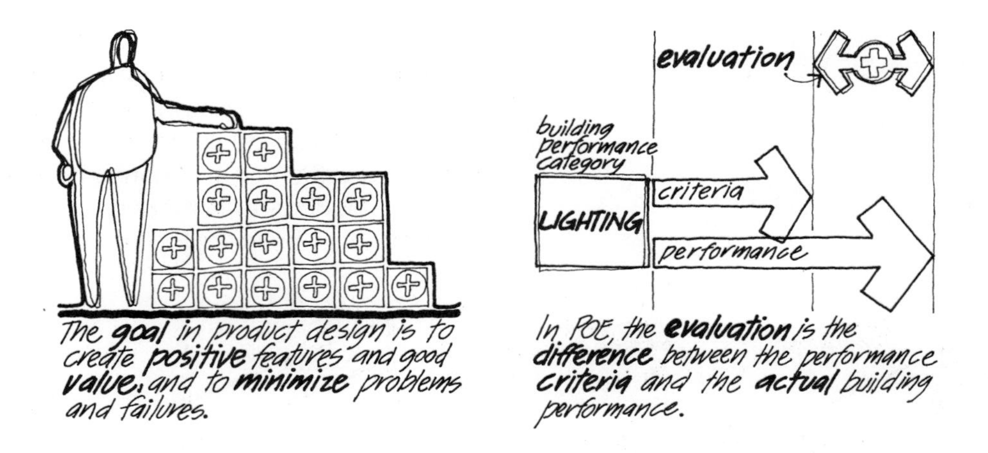

*The **goal** in product design is to create **positive** features and good **value**; and to **minimize** problems and failures.*

*In POE, the **evaluation** is the **difference** between the performance **criteria** and the **actual** building performance.*

By analogy, POEs are intended to compare systematically and rigorously the actual performance of buildings with explicitly stated performance criteria; the differences between the two constitute the evaluation.

USES AND BENEFITS OF POE

Depending on the objectives of the client organization and the time frame involved, POEs have uses and benefits over the short, medium, and long term (table 1-1).

*POE **benefits** may be **short**-term, **medium**-term or **long**-term.*

SHORT-TERM BENEFITS
- Identification of and solutions to problems in facilities
- Proactive facility management responsive to building user values
- Improved space utilization and feedback on building performance
- Improved attitude of building occupants through active involvement in the evaluation process
- Understanding of the performance implications of changes dictated by budget cuts
- Informed decision making and better understanding of consequences of design

MEDIUM-TERM BENEFITS
- Built-in capability for facility adaptation to organizational change and growth over time, including recycling of facilities into new uses
- Significant cost savings in the building process and throughout the building life cycle
- Accountability for building performance by design professionals and owners

LONG-TERM BENEFITS
- Long-term improvements in building performance
- Improvement of design databases, standards, criteria, and guidance literature
- Improved measurement of building performance through quantification

Table 1-1. Post-occupancy evaluation uses and benefits.

Short-term Benefits

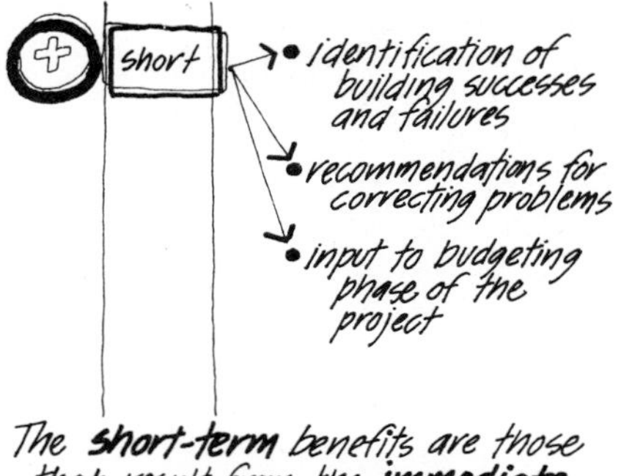

The **short-term** benefits are those that result from the **immediate** use of POE findings.

Over the short term, successes and failures in the performance of buildings are identified and recommendations made for the appropriate action required to resolve any problems. Additional study may be needed to understand the identified problems fully, in which case further in-depth POEs may be undertaken.

Another short-term benefit of POE pertains to the budget cutting that is common in the fiscal-planning phase of the building process. Reducing a project's cost often results in inferior quality, which in turn can negatively affect the functioning of the organization occupying the building. POEs can help to show the implications of various design alternatives devised to meet lowered budgets, enabling the achievement of the best level of quality and performance within these constraints.

Medium-term Benefits

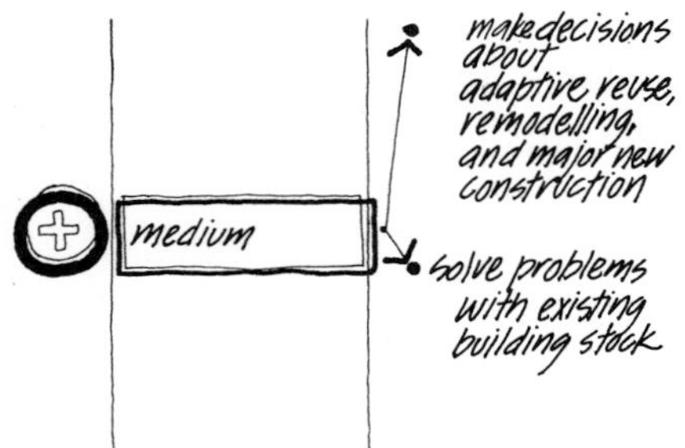

Medium-term POE benefits relate to **major** decisions about **building construction.**

Over the medium term, POEs can provide the justification and information base for adaptive reuse, remodeling, or major construction in order to resolve problems that have been identified in existing buildings. Recycling old loft buildings into apartments, installing new telecommunications wiring, or building additions to accommodate organizations' changing space needs are examples.

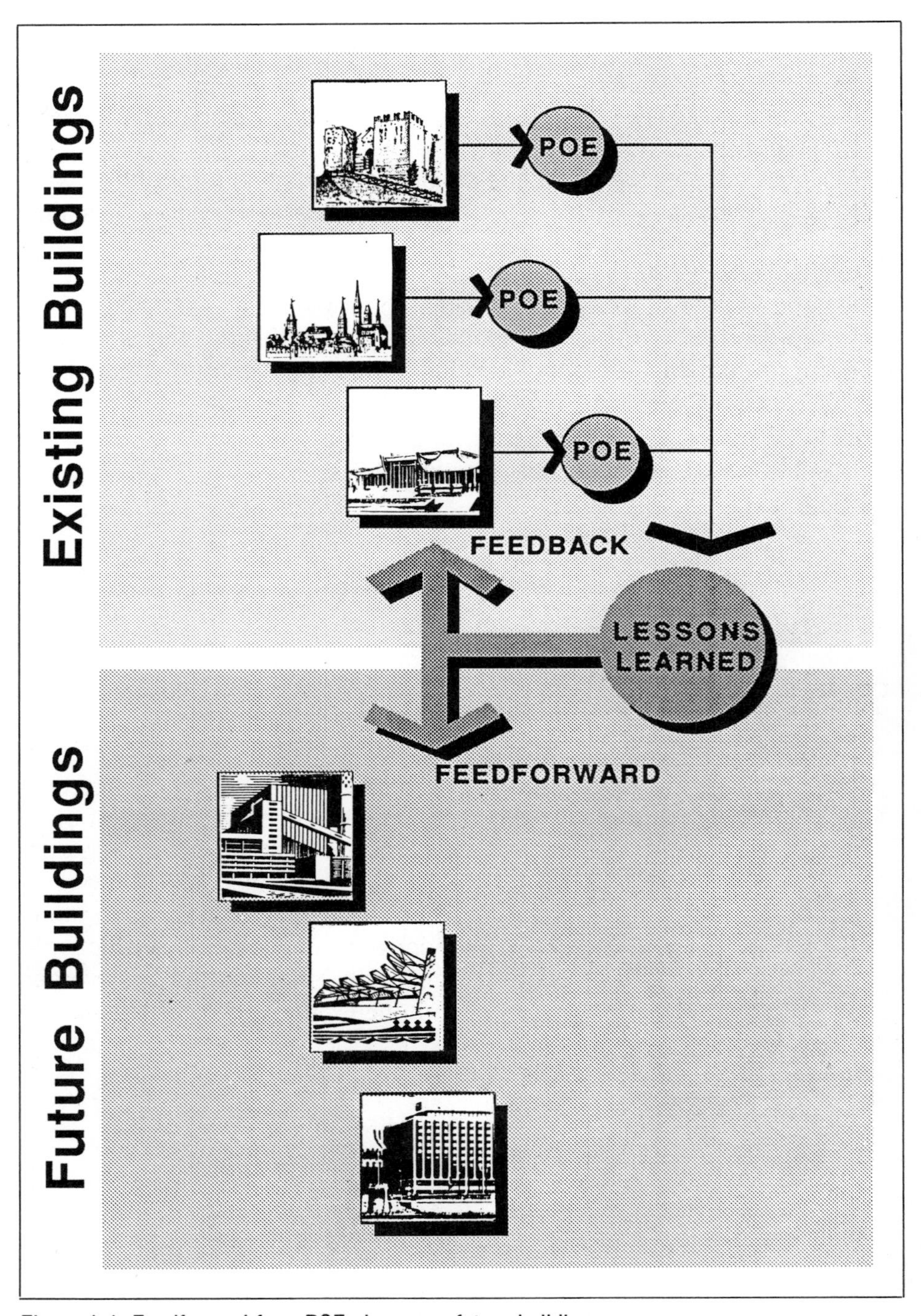

Figure 1-1. Feedforward from POEs improves future buildings.

Long-term Benefits

Long-term benefits result when the lessons learned from the failures and successes of building performance are applied to the design of future buildings (fig. 1-1).

The time frame for long-term benefits to come to fruition can range from three to ten years. As was stated earlier, this benefit of POEs is particularly relevant to generic building types, such as hotels, office buildings, schools, retail facilities, and housing.

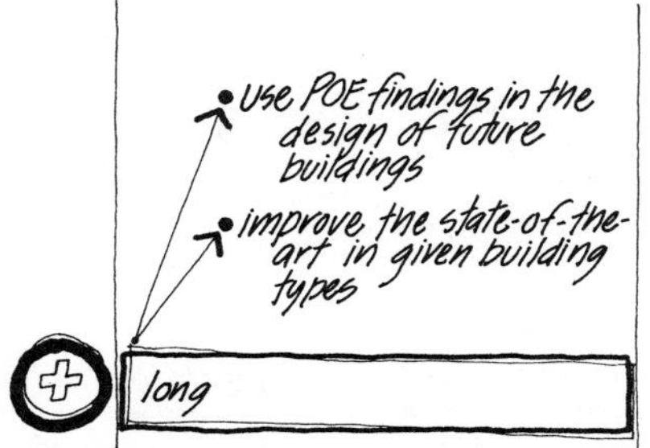

The **long-term** benefits involve the **sharing** and **feeding forward** of POE results for use in the building industry as a **whole.**

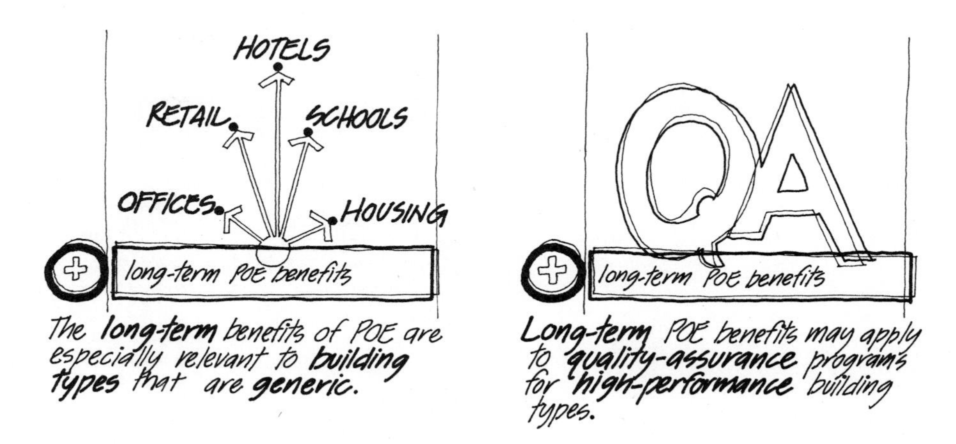

The **long-term** benefits of POE are especially relevant to **building types** that are **generic.**

Long-term POE benefits may apply to **quality-assurance** programs for **high-performance** building types.

Another long-term benefit of POE is its application to quality assurance (QA). This area of concern is of increasing importance in the medical field, where performance standards have been established, for instance, in the area of laboratory testing. These standards provide for the appropriate levels of accuracy of testing procedures and must be carefully followed. Quality assurance requires some means of indicating quality and performance, action taken on problems identified by these indicators, and accurate recordkeeping. QA, adapted to the building industry, can be implemented by employing the POE process model presented in chapter 4.

As outlined in table 1-1, POEs can result in a broad range of recommendations and actions. Although POEs are a rational and easily understood activity in the building process, some professionals still believe that such evaluations have the potential to undermine the integrity of design and planning, because in our litigious society, POE can provide evidence in lawsuits where malpractice in architecture and planning is suspected. As appendix A, on legal issues in POE, indicates, these are serious concerns; however, they should not obviate the fact that POEs provide significant benefits to architects, building managers, and client organizations alike.

Some professionals still believe that POEs may **threaten** the **integrity** of the design discipline.

THE HISTORY OF POE

Informal and subjective building evaluations have been conducted throughout history. Systematic POEs, as required for today's complex buildings, however, employ explicitly stated performance criteria with which performance measures of buildings are compared. Today, this type of evaluation is carried out as a routine activity in both the public and private sectors.

Informal *and* ***subjective*** *evaluations of buildings have existed throughout history.*

Today, POE is done ***rigorously*** *by both the* ***public*** *and* ***private*** *sectors.*

POE is said to derive its name from the occupancy permit that is issued when a building is completed, inspected, and deemed safe according to building codes and regulations (Bechtel 1980). The first significant efforts at POE were made in the mid-1960s when severe problems, some of which were attributable to the built environment, were observed in institutions such as mental hospitals and prisons (Osmond 1966). The interest in evaluating the health, safety, security, and psychological effects of buildings on their occupants was further stimulated by Robert Sommer's books *Personal Space: The Behavioral Basis of Design* (1969) and *Tight Spaces: Hard Architecture and How to Humanize It* (1974), as well as by Edward T. Hall, who wrote *The Hidden Dimension* (1966) and *The Fourth Dimension in Architecture: The Impact of Building on Man's Behavior* (1975).

The 1960s saw the growth of research focusing on the relationships between human behavior and building design, which led to the creation of the new field of environmental design research and the formation of interdisciplinary professional associations, such as the Environmental Design Research Association in 1968. Members of these associations include architects, planners,

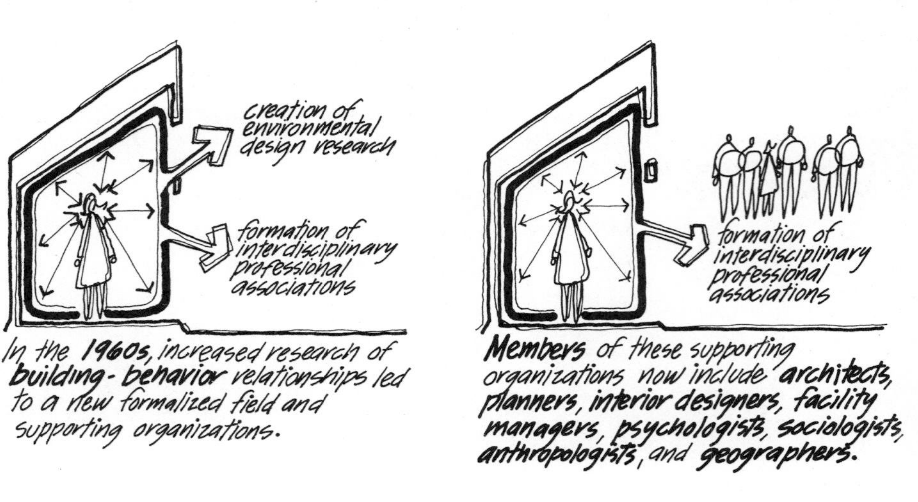

In the ***1960s****, increased research of* ***building-behavior*** *relationships led to a new formalized field and supporting organizations.*

Members *of these supporting organizations now include* ***architects, planners, interior designers, facility managers, psychologists, sociologists, anthropologists****, and* ***geographers.***

facility managers, interior designers, psychologists, sociologists, anthropologists, and geographers. Emerging professional specializations include human behavior-based building research, facility programming, and, most important, POE. Further manifestations were the increasing numbers of publications (Preiser 1978), including books on evaluation methods, case study applications, and new journals that featured POEs. *Environment and Behavior,* the *Journal of Environmental Psychology,* and major architectural magazines, such as *Architecture* and *Progressive Architecture,* began to publish building evaluations.

*Post-occupancy evaluation has been **reinforced** by proponents of a more **rational** design process.*

The emergence of POE techniques and the publication of evaluation studies were reinforced by proponents of more rational and rigorous design processes in architecture. Christopher Alexander, an early leader in this field, wrote three influential books: *Notes on the Synthesis of Form* (1964), *Houses Generated by Patterns* (1969), and *A Pattern Language* (1977). These publications introduced the notion of design requirements and patterns into the design process, based upon the evaluation of the needs of those for whom the designs were intended. These efforts set the stage for government agencies, such as the General Services Administration (1975), to begin research on measurable performance criteria for office buildings. Their contribution to the field is but one of the many milestones in the evolution of POE (table 1-2). These milestones fall into three categories corresponding to the 1960s, 1970s, and 1980s respectively.

YEAR	*AUTHOR(S)*	*BUILDING TYPE(S)*	*CONTRIBUTION TO THE FIELD*
1967	Van der Ryn and Silverstein	Student dormitories	Environmental analysis concepts and methods
1969	Preiser	Student dormitories	Environmental performance profiles; correlation of subjective and objective performance measures
1971	Field	Hospital	Multimethod approach to data collection
1972	Markus et al.	Any facility type	Cost-based building-performance evaluation model
1974	Becker	Public housing	Cross-sectional, comparative approach to data collection and analysis
1975	Francescato et al.	Public housing	Evaluation models of resident satisfaction, allowing physical managerial intervention.
1975	General Services Administration	Office buldings	Office-systems performance standards

Table 1-2. Milestones in the evolution of POE. *(continued)*

YEAR	AUTHOR(S)	BUILDING TYPE(S)	CONTRIBUTION TO THE FIELD
1976	U.S. Army Corps of Engineers	Military facilities	Design guide series with updatable, state-of-the-art criteria
1976	Rabinowitz	Elementary schools	Comprehensive, full-scale evaluation of technical, functional, and behavioral factors
1979	Public Works Canada	Government facilities	POE incorporated into project delivery system
1980	Daish et al.	Military facilities	POE as routine staff activity in government building process
1981	Marans and Spreckelmeyer	Offices	Evaluation model linking perceptual and objective attributes
1982	Parshall and Peña	Any facility type	Simplified and standardized evaluation methodology
1983	Orbit I	Offices	Office research linking buildings and information technology
1984	Brill et al.	Offices	Linking worker productivity and office design
1985	White	Any facility type	Linking programming and POE in graduate architectural education
1986	Kantrowitz et al.	Architecture school	POE analysis of entire building process and documentation
1986	Preiser and Pugh	Any facility type	POE process model and levels of effort

Table 1-2. (Continued)

The 1960s: Institutional Settings

Early POE efforts were heavily biased toward college dormitory evaluations, primarily because of their ready availability to university researchers and generally cooperative dormitory residents. Van der Ryn and Silverstein's POE on dorms at Berkeley (1967) led to similar evaluations by Hsia (1967) and Preiser (1969). These studies yielded startling findings, such as significant inefficiencies, misfits between users and buildings, and a stronger than anticipated connection between building configuration and the formation of social relationships.

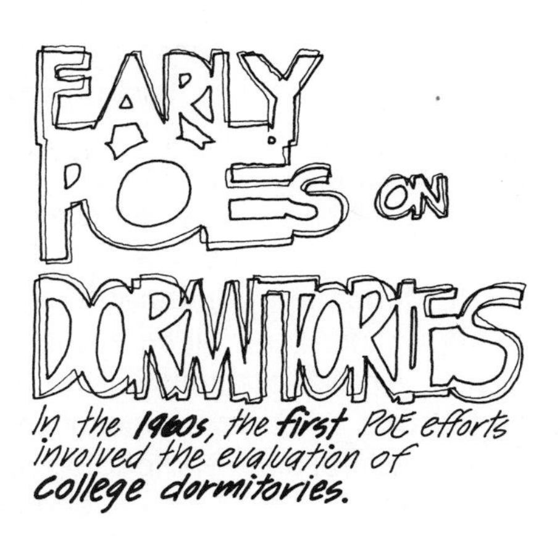

*In the **1960s**, the **first** POE efforts involved the evaluation of **college dormitories.***

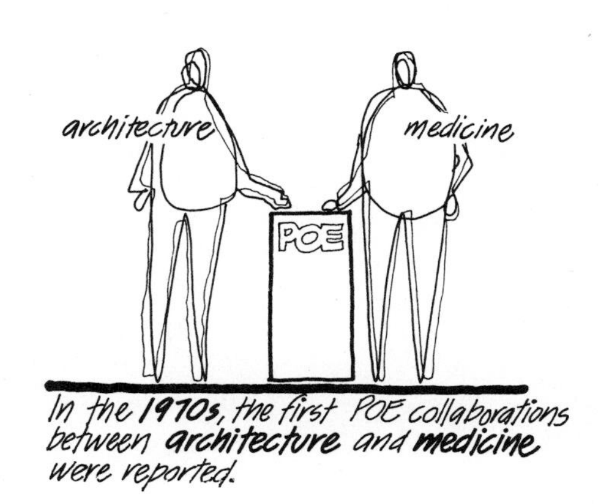

*In the **1970s**, the first POE collaborations between **architecture** and **medicine** were reported.*

Much of the research work carried out in the late 1960s was published in the early 1970s, when the first collaborations between architectural and medical professionals were reported, especially in the area of hospital design. For example, Field (1971) conducted a comprehensive building evaluation of the Tufts University Medical Center, addressing such variables as staff travel time, proximity of hospital functions, time actually spent at patients' bedsides, differences in staffing requirements and efficiency. A similar effort was undertaken by Trites et al. (1970). Concurrently, pioneering POE efforts in Great Britain by the Pilkington Research Unit (Manning 1965; Canter 1970) reported on building evaluation of offices and schools.

The 1970s: Systematic, Multimethod POEs

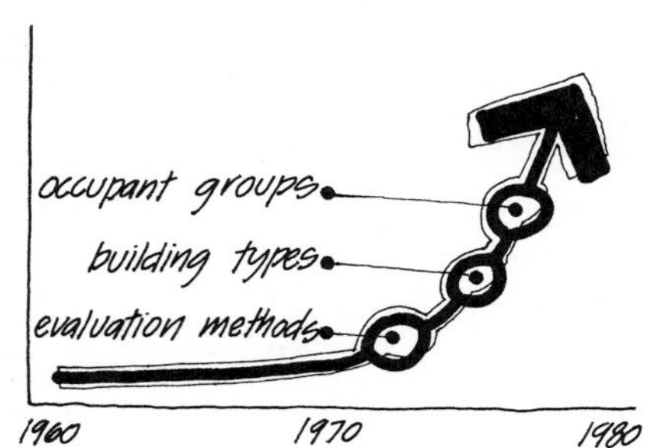

*POE activity increased **dramatically** in the **1970s** with the emergence of new knowledge about POE methods, building performance, and building users.*

The 1970s witnessed a dramatic increase in the use of POEs, as a considerable body of knowledge emerged on evaluation methods, building types, and occupant groups (Preiser and Daish 1983).

Perhaps the first attempt at systematic building evaluation was made by Markus and his colleagues at the Building Performance Research Unit (BPRU) at the University of Strathclyde, Scotland, in the late 1960s as described in *Building Performance* (Markus 1972). Markus proposed a cost-based building evaluation model that described the interacting elements of building systems, environmental systems, and activity systems, as well as the ultimate goals and objectives of owners and occupants to be achieved in building performance.

Oscar Newman's research on crime in high-rise public housing, in his book *Defensible Space* (1973), was also a critical influence at that time. His study highlighted the relationships between the occurrence of crime and project size, scale, layout, and ability to control territorial spaces in public housing. This work not only changed housing policy in terms of design in the United States, but it also underscored the power of POE methods and their resulting benefits.

Other milestone evaluations of housing were carried out by Clare Cooper, who emphasized the application of survey, interview, and observation techniques in POE data collection, for example, in the evaluation of St. Francis Square (Cooper 1970) and Easter Hill Village (Cooper 1975) in San Francisco. These single-project POEs carefully addressed many specific design features that contributed to the successes or failures of the evaluated housing schemes, such as the use of landscaping, provision of vehicular-free pedestrian zones, functionality and flexibility of floor plans of the housing units, quality of construction, adequacy of storage, and the use of color.

Becker (1974) conducted an evaluation of a number of multifamily housing projects sponsored by the New York Urban Development Corporation. This POE generated comparative data that were used to identify patterns within and across housing types and user groups, as well as suburban and urban sites. A major contribution was the use of triangulation or data-collection methods such as surveys, interviews, systematic observations, behavioral mapping, archival data, and photographic records. This POE was one of the first to evaluate the effect of management on housing developments and their physical characteristics.

Another important POE project of the mid-1970s significantly influenced policy of the U.S. Department of Housing and Urban Development (Francescato et al. 1979; Weidemann et al. 1982; Francescato et al. in press). This project tested the nature and relative importance of various factors that contribute to housing residents' satisfaction. A set of reliable techniques for evaluating housing satisfaction was developed that has since been used by other researchers. This POE permitted the involvement of residents in the improvement of their housing and helped in selectively directing limited modernization resources to those aspects of housing development that were most likely to increase resident satisfaction.

Other government organizations, such as the U.S. Army Corps of Engineers, sponsored a pioneering series of evaluation projects on building types ranging from U.S. Army service schools (Department of the Army 1976) to recreation centers, the results of which became the basis for a series of design guides. In this instance, the results of systematic evaluation research were developed into design criteria and guideline documents created to assure future better-quality buildings.

Goodrich (1976) conducted a "post-design evaluation" that examined the response of six hundred employees to open office planning and innovative task lighting in a newly opened office building. This work was one of the first POEs in the field of office design. Psychological and organizational measures enabled Goodrich to gauge the employees' satisfaction with the performance of their office environment.

Rabinowitz's *Buildings in Use Study* (1975) evaluated four schools in Columbus, Indiana, all of which had been designed by nationally prominent architects. This POE was innovative in its comprehensiveness and depth. It examined the technical, functional, and behavioral elements of performance in schools in great detail and is reported in chapter 10.

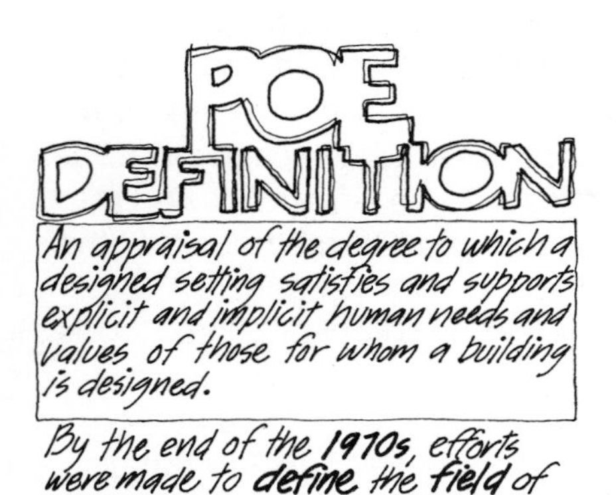

By the end of the 1970s, the first books on POE were published and attempts were made at defining the field of POE as: "an appraisal of the degree to which a designed setting satisfies and supports explicit and implicit human needs and values" of those for whom a building is designed (Friedman et al. 1978, p. 20).

In their more rigorous forms, POEs involve a careful and systematic approach: each user group must be represented and each important design element must be examined. However, as stated by Friedman et al. (1978, p. 2), "Individual judgments by the designer and the layperson must not be abandoned,

*In a **rigorous** POE, all **user groups** are represented and all important **building elements** are studied.*

but must be augmented by more complete and rigorous techniques." The inclusion of all relevant elements means the consideration, for example, of health, safety and security, functional requirements, lighting, acoustics, thermal needs, and comfort, as well as qualitative concerns that affect the psychological comfort and satisfaction of building occupants. Through the 1970s, however, most approaches focused primarily on user satisfaction—and references to the attributes of the actual physical environment were rarely made. As a result, the ability to generalize from the findings of those POEs was limited.

The social-science-based approach to design evaluation by Friedman et al. (1978) presented a much more comprehensive perspective. This POE framework included the setting, clients, proximate environmental context, design process, and the social/historical context. Evaluation included a definition of the larger context within which design occurs, the selection of methods, analysis of results, the definition of the focus of the evaluation, and the feedback of the evaluation information into the building process.

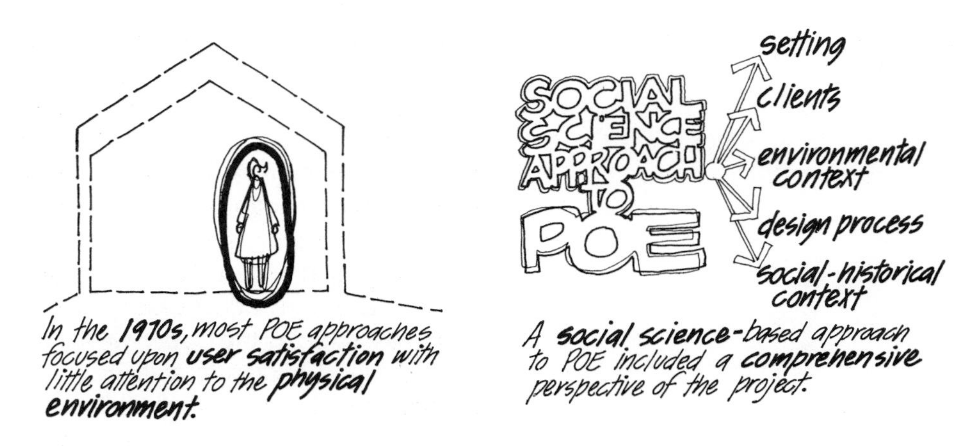

*In the **1970s**, most POE approaches focused upon **user satisfaction** with little attention to the **physical environment**.*

*A **social science**-based approach to POE included a **comprehensive** perspective of the project.*

The 1980s: POE Practice in the Public and Private Sectors

*In the **1980s**, POE has developed into a discipline of its own with **standardized terms**, **practitioner networks**, and **large multibuilding** evaluations.*

In the 1980s, POE developed into a discipline of its own. Standardized terms are now being used in POE (see the glossary), networks of practitioners and researchers have been formed, and significant large-scale multibuilding POEs are being carried out.

The 1980s have produced a number of advances in theory, method, strategy, and application of POE. An important model of how the physical environment and organizational setting of the workplace influence the perceptions and behavior of workers was developed by Marans and Spreckelmeyer in *Evaluating Built Environments: A Behavioral Approach* (1981). Their conceptual model relates objective environmental attributes to subjective user perceptions and assessments of the effect of the work environment on occupant behavior and overall satisfaction. The model suggests that an individual's expressed satisfaction with a building is dependent upon the assessment of the physical and organizational attributes of the building. In turn, this is dependent upon how the building is perceived, as well as on the standards or criteria against which it is judged. The model further highlights the need for objective measures of the physical environment that can be obtained without surveys—for example, through supervisor's ratings, analysis of records, archival research, direct observations, and so on.

Another POE approach was described by Preiser (1983) with an emphasis on the use of explicitly stated performance and evaluation criteria on the one hand, and a correlation between subjective and objective measures of performance on the other.

A different POE strategy was put forth by Parshall and Peña in *Evaluating Facilities: A Practical Approach to Post-Occupancy Evaluation* (1982). This evaluation strategy is closely linked to their widely known programming approach called problem seeking (Peña et al. 1977), the major elements of which are function, form, economy, and time. These four elements are considered throughout the POE process of inquiry, using four steps: establishment of purpose; selection and analysis of quantitative information; identification and examination of qualitative information; and statement of the lessons learned from the evaluation.

A pioneering POE in terms of magnitude was conducted by Brill et al. (1984). The results of this five-year, multisponsor research effort were based on an evaluation that ultimately included seventy offices and some five thousand workers. The primary method of data collection was an extensive questionnaire combined with physical measurements of office environments. The study isolated relationships among specific physical factors, job satisfaction, and performance, as well as ease of communication in the workplace.

Beginning in the late 1970s and continuing into the 1980s, several projects were initiated that incorporated POE into the building process of government agencies, such as the public works departments in both Canada (1979) and New Zealand (Daish et al. 1980, 1981). The intent in New Zealand was to introduce POE as a routine staff function within government agencies, in this case the military, with the idea that POEs were to optimize space utilization and identify needed improvements in buildings.

The POE process model presented in chapter 4 evolved over a period of five years and constitutes the authors' cumulative experience in having conducted numerous POEs, both in the academic realm (Eshelman, Preiser, et al. 1981) and commercially as architectural research consultants (Preiser and Pugh 1986a,b; Preiser 1986).

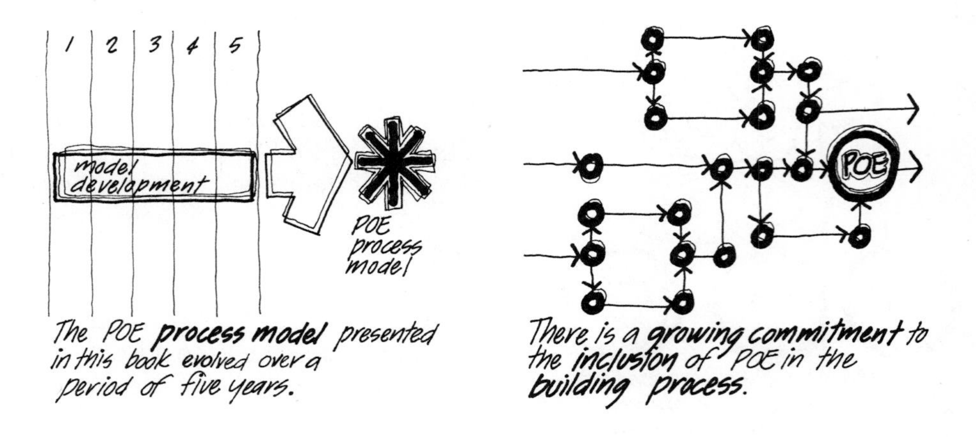

The POE **process model** presented in this book evolved over a period of five years.

There is a **growing commitment** to the **inclusion** of POE in the **building process**.

While POE is still a maturing field, there appears to be a growing commitment toward the inclusion of POE in the building process, just as the activity of programming has been accepted as one of the critical steps in the predesign phase of the building process (Building Research Board 1986, 1987).

Another sign of the maturing of the POE field is the availability of specialized courses on the subject matter in both undergraduate and graduate curricula at schools of architecture in the United States. The first master's degree option offering specialization in POE and facility programming was recently organized at Florida A & M University (White 1985).

The foundation of the Environmental Design Research Association (EDRA) in 1968 was mentioned earlier. The growth of POE, which is a major part of the design research field, is indicated by the recent formation of three new, relevant international organizations in different parts of the world:

- International Association for the Study of People and Their Physical Surroundings (IAPS) is based in Europe.
- People and the Physical Environment Research Association (PAPER) is based in Australia, New Zealand, and Southeast Asia.
- Man–Environment Research Association (MERA) is a Japanese organization.

2. The Profession and POE

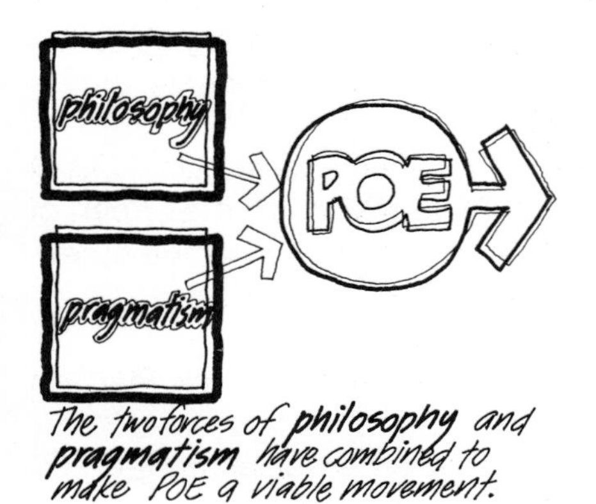

The two forces of ***philosophy*** *and* ***pragmatism*** *have combined to make POE a viable movement.*

Two forces, philosophy and pragmatism, have converged to make POE a viable "movement." The previous chapter dealt with POE's philosophical basis, as well as the growth of knowledge and improvements in methodology. This chapter focuses on critical pragmatic issues that involve architects and members of the building industry.

BUILDING EVALUATION AND PRACTICE

The results of building evaluation and feedback have been used for centuries, particularly after there has been a major building failure. These evaluations have resulted in regulations that historically have often been the only systematic and research-based source of information on building design. Regulations evolved into building codes, which began to control critical aspects of buildings, such as the health, safety, and general welfare of building occu-

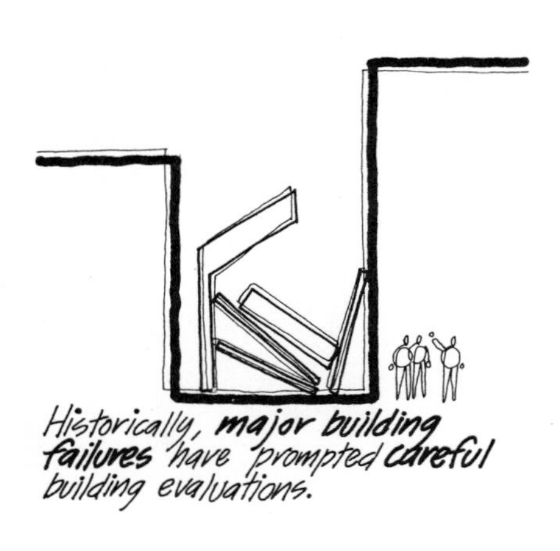

Historically, ***major building failures*** *have prompted* ***careful*** *building evaluations.*

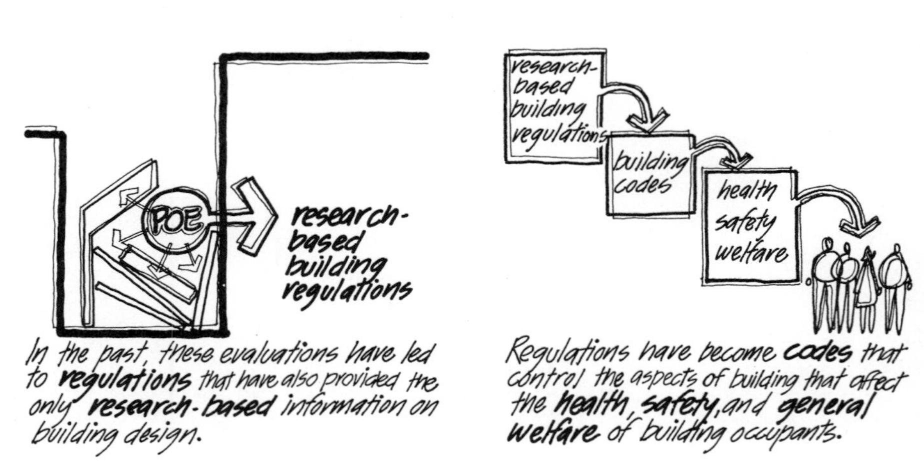

In the past, these evaluations have led to ***regulations*** *that have also provided the only* ***research-based*** *information on building design.*

Regulations have become ***codes*** *that control the aspects of building that affect the* ***health, safety,*** *and* ***general welfare*** *of building occupants.*

pants. Over the years, new building types emerged, construction grew more complex, and additional aspects of building design were codified. Finally, when psychological and sociological considerations were linked to design, the study of environment and human behavior became a new discipline and knowledge from this discipline was also applied to building evaluations. Taking these developments into consideration, three elements of building perfor-

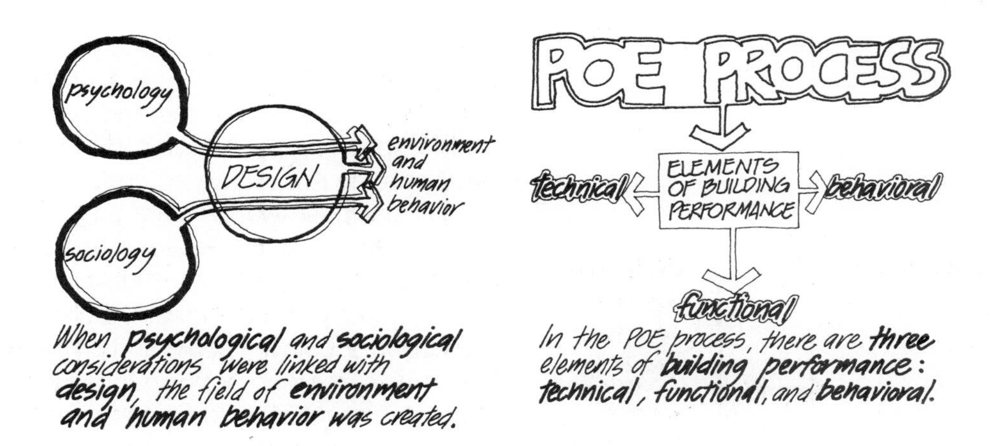

mance can be identified and treated hierarchically in their application to the POE process:

1. *Technical elements:* Health, safety, and security aspects of building occupancy
2. *Functional elements:* Occupants' ability to operate efficiently and effectively
3. *Behavioral elements:* Psychological and social aspects of user satisfaction and general well-being

Technical elements of POE include basic survival issues such as fire safety, structural integrity, and sanitation, as well as other factors pertaining to durability, acoustics, and lighting. Although technical building evaluation has been developed and used for centuries, recent changes in building products, building size and complexity, and the relationships among the participants in the building process have stimulated the need for POE.

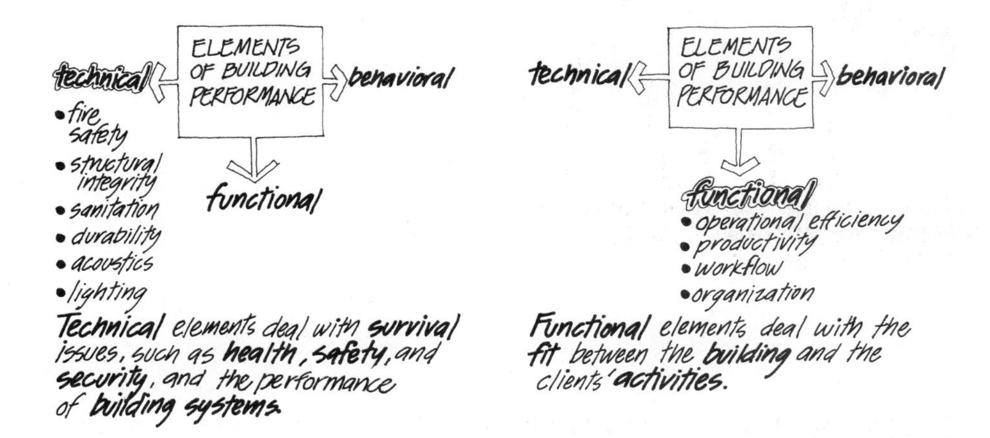

Functional elements of POE began to be treated formally with the emergence of a variety of new building types in the nineteenth century. Guidebooks covered the specialized functions in the design of greenhouses, prisons, hospitals, department stores, and housing, among others. This trend has continued unabated, accelerated by such factors as the proliferation of building types; new and constantly evolving building systems; new technologies; and new processes and methods within organizations. The amount of detail that is specified by and can be evaluated for each of these elements of performance has increased considerably.

Over the past twenty years as POEs have developed, a critical mass of expertise, findings, applications, and credibility has accumulated. Issues such as privacy, security, the symbolism of buildings, social interactions, perceptions of density, and territoriality are included in a POE as behavioral elements. Furthermore, both the public and private sectors have recognized that behavioral elements of building performance factors had been overlooked for decades, to the detriment of building occupants and owners alike.

The three emphases of building evaluation mentioned above, technical, functional, and behavioral, are the focus both of the philosophical and methodological advances described in chapter 1 and of the pragmatic issues discussed here.

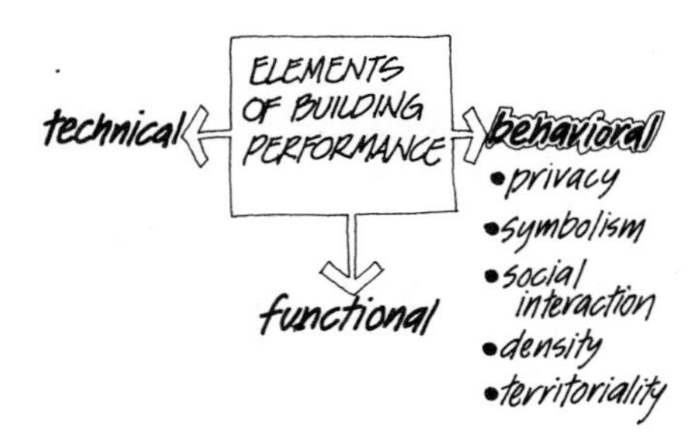

Behavioral elements deal with the perceptions and psychological needs of the building users and how these interact with the facility.

A REVIEW OF BUILDING PERFORMANCE

How well do buildings work? Is there really a need for POE? Has there ever been any comprehensive examination of the performance of buildings in use? Where does responsibility lie for flaws in the finished product? What are the consequences of architects' design decisions, both positive and negative?

Very little has been done in a systematic and rigorous way to explore the performance of a large sample of buildings. In probing "building demographics," it is necessary to examine a variety of partial studies and circumstantial evidence from a number of sources.

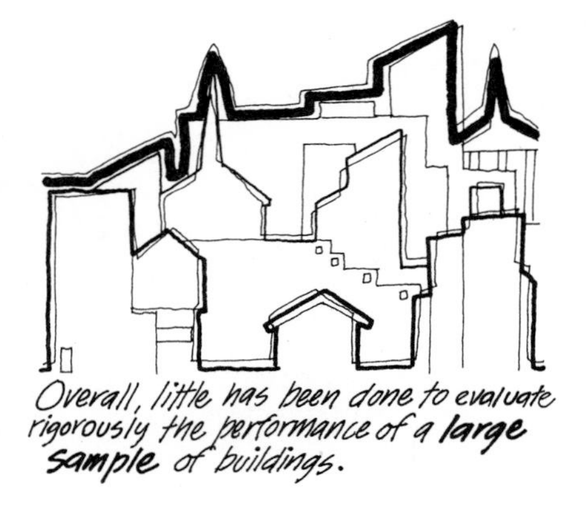

Overall, little has been done to evaluate rigorously the performance of a large sample of buildings.

Building Performance in Research Findings

In the United Kingdom, a research project that attempted to assess the technical quality of a large number of buildings was conducted by the Building Research Establishment (BRE), assisted by the National Building Agency (Bonshor and Harrison 1982). Even though this study, completed late in 1981, concentrated on one building type in terms of technical adequacy prior to occupancy, it did examine fifteen housing projects with a total of seven hundred seventy-nine units. It was performed by competent government building-research agencies and resulted in a highly detailed evaluation. The buildings were evaluated against criteria documented in existing building regulations and standards.

An average of twenty-eight faults per housing unit was found. Half of the faults had their origin in design or specifications, and the other half were site-related. A follow-up study provided the results to eight architectural firms designing similar projects. Using these data, the architects found twenty-five faults per dwelling in their own plans and were able to correct them prior to construction.

Unfortunately, this significant study, although not a POE, is unique. It indicates that even in a regulated and technical context, the realm of building codes and standards, the context most familiar to architects, there should be serious concern for building performance.

Another study, also not a POE, supports part of the BRE study results. In an examination of four hundred seventy-seven construction claims in twenty-two U.S. government projects, the researchers found that "fully 46 percent of additive claims were due to design errors." (Diekmann and Nelson 1985). Geoffrey Scott documented large numbers of structural failures caused by the lack of quality control when certain additives are used in concrete mixes. More than 50,000 buildings in Britain used these additives until the practice was halted in the 1970's (Scott 1976).

Serious **design errors** can occur, even in a highly **regulated** building situation.

Chapter 1 highlighted milestones in the development of POE, significant projects that were systematic, rigorous, and provided important advances in methodology and knowledge about building performance evaluation. Many of these projects emphasized particular aspects of POE, for instance, security, communication, or another specific functional aspect of building performance. Taken together, however, the projects provide a spectrum of results that evaluated existing environments and implicitly measured "how well buildings work."

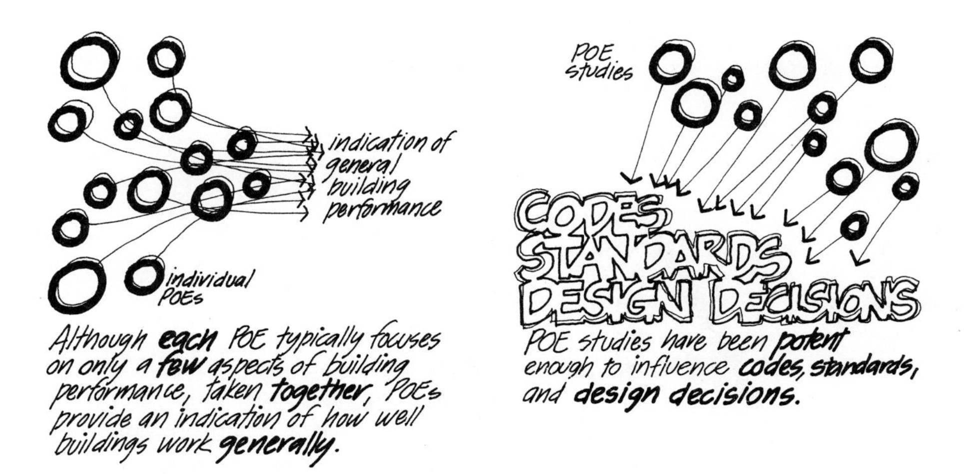

Although **each** POE typically focuses on only a **few** aspects of building performance, taken **together**, POEs provide an indication of how well buildings work **generally**.

POE studies have been **potent** enough to influence **codes, standards,** and **design decisions.**

All of the milestone POEs listed in table 1-2 found the buildings they evaluated wanting in many areas. Altogether, literally hundreds of critical flaws were identified. Some of these POEs, individually or as part of a larger movement, were potent enough to have had a definitive influence on building standards and design.

Dorms at Berkeley (van der Ryn and Silverstein 1967) influenced the creation of a new generation of dormitory designs for state universities in California and the rest of the United States. Another key investigation, that of Newman (1973) on security in public housing, helped change federal policy for that building type and provided innovative design concepts for new and existing housing projects, many of which have since been renovated. The Buffalo Organization for Social and Technical Innovation (BOSTI) studied office buildings on a large scale using data from a national cross section of corporations. The study had a significant influence on office design, partly reinforcing both positive and negative aspects of open office design concepts (Brill et al. 1984).

Other notable examples indicate how little was known about hospitals, buildings that are very costly and whose design can be critical to the life of their patients. Two hospital-focused studies were conducted by McLaughlin et al. (1972) and Clipson and Wehrer (1973). McLaughlin and his coevaluators examined the frequency and intensity of change in different parts of a hospital. Their study found that radiology and laboratory segments underwent frequent renovation both minor and major. When combined with the cost implications of these changes, the evidence demonstrated that the use of interstitial service/structural floors in such places as hospitals and laboratories is economically and functionally beneficial. The Clipson and Wehrer study of cardiac-care facilities in hospitals documented the obstacles and inefficiencies, many due to design considerations, that cost precious seconds in administering emergency cardiac care.

Reviews are sometimes a good source of information about how well buildings work.

Literature reviews of a single building type or attribute are a potential source of information on how well buildings work. For example, Ahrentzen et al. (1982) and Weinstein (1979) examined 110 and 144 studies respectively concerning behavioral elements of performance in school environments. However, key variables such as classroom student density, size, shape, and seating arrangements seem not to have been examined at all. Among other issues, the location and size of windows, aspects of small group behavior, and privacy also need to be examined.

The Crisis Model

The crisis model of building evaluation illustrates both the POE process and areas of continuing problems in building performance.

The "crisis model" of building performance illustrates not only an evaluation process in action, but also the extent of continuing problems in building performance. It answers the question: How well do buildings work, not in theory but in practice?

It often takes tragedy to instigate reform. It was the Fire of London in 1666 that generated the London Building Codes, specifying criteria for street widths, building heights, external materials (brick), and window areas—all elements that were to become important determinants in the design of Georgian London. The New York tenement housing codes of 1879 and 1901 contained criteria for light, air, and sanitation as a reaction to the terrible living conditions and general squalor at that time. The Our Lady of the Sorrows School fire in Chicago in 1958, in which ninety-two students and three teachers lost their lives, produced code changes that introduced the concept of building compartmentalization with fire walls that would prevent the spread of smoke. Incidents of high-rise fires as recently as the 1970s have resulted in code requirements for sprinklers and positive air pressure in stairwells in many buildings. The 1980 Las Vegas MGM Grand Hotel fire, in which eighty-four persons died, resulted in further changes in hotel fire safety and codes.

Tragic events such as these, and the evaluation of their causes, are part of a continuing process in an industry that is fragmented and highly complex and has seen little systematic evaluation and feedback in the past. It should be recognized that the health, safety, and welfare aspects are only a small part of total building performance, albeit a crucial and highly regulated one. These aspects of building performance are also the most developed in terms of methodology and knowledge base, as well as having an established infrastructure of organizations, testing laboratories, publications, and the like.

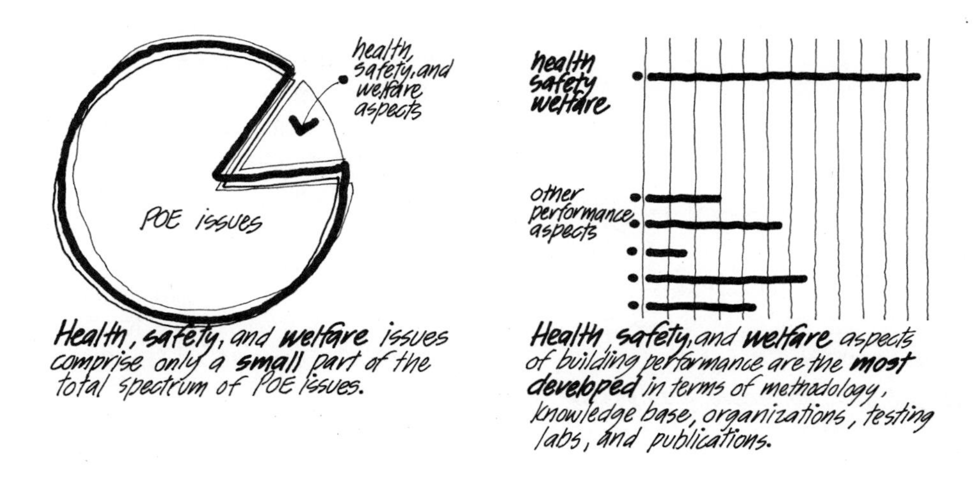

Health, safety, and welfare issues comprise only a small part of the total spectrum of POE issues.

Health, safety, and welfare aspects of building performance are the most developed in terms of methodology, knowledge base, organizations, testing labs, and publications.

An apt illustration of effects on building performance can be seen in responses to the energy crisis of the 1970s that generated many strategies and criteria for energy conservation. Reducing the rate of air infiltration and heat loss in buildings was universally accepted as a part of any overall design solution for buildings in the northern United States. Increased amounts of insulation material, wrapping buildings with airtight and watertight membranes, additional caulking and sealants, and reduced outside air exchange rates became commonplace. Underground and earth-sheltered buildings, although infrequently used, were highly publicized. How well did these strategies really work in response to the new performance criteria?

A common tactic was to retrofit with highly insulative materials such as urea-formaldehyde foam. Some 500,000 homes had this product installed in the 1970s. In 1982, after years of testimony and investigation concerning the safety of the product by the Consumer Product Safety Commission, and after a considerable drop in its use because of the resulting publicity, urea-formaldehyde was banned in buildings. Reported health hazards from the fumes given off by this material included respiratory irritation, coughing, and nausea. Rats, subjected to the same levels of formaldehyde that humans were exposed to, developed cancer as well as mutations and chromosomal damage.

There is no doubt that the effects of this material were concentrated by the "tight house" concept that resulted from the use of so-called foamed-in-place insulation. Formaldehyde fumes, emitted by plywood and particle board, are among a number of pollutants in tight buildings. Recent changes in European building codes include a mandated period for the emission of gas from construction materials after construction has been completed and before occupancy can take place. This allows the chemicals used in building to evaporate and be eliminated from the environment. Other major studies of the effects of tight environments also show increased indoor pollutant levels. Even indoor pollutants from the ubiquitous gas range were higher than recommended outdoor levels in tight houses during winter months.

This type of phenomenon has been reported frequently enough to have received its own name: the tight-building syndrome (TBS). Incidents of problems in tight buildings include pollution from substances in construction materials, both those already mentioned and paints, preservatives, waxes, and others, from outside air pollutants, such as automobile fumes that enter a building's air-intake system, and from interior pollutants such as cigarette smoke, fumes from cleaners and aerosols, and even biological contamination.

A third consequence of the tight-building strategy concerns the components of the building itself. When buildings are tightened excessively, moisture can be trapped in walls and roofs. Trapped moisture hastens deterioration, reduces insulative qualities, and in time affects the wall surfaces. Because these consequences cannot be seen, they are particularly insidious.

- innovation
- building-component and system linkages
- failure to employ POE

Many building failures have happened because of **premature innovation**, *failure to understand relationships between building* **systems**, *and reluctance to use* **POE**.

How well do buildings work? The illustrations discussed here document only a few of the consequences of strategies that were meant to help solve the effects of the energy crisis. In the 1970s, at a time when there was much expertise, infrastructure, and regulation in this area, many such failures occurred. Primarily, however, the failures happened because the building industry adopted innovation too readily, the linkages between various building components and systems were not well understood, and POE was not used.

Are the energy-related problems an isolated phenomenon? Unfortunately, there is evidence of other short-, medium-, and long-term building-related problems that have had serious and widespread effects. The consequences of exposure to asbestos, a material commonly used for many decades in buildings, are well understood (Craighead and Mossman 1982). Scott (1976) documented structural failures due to the lack of quality control in the use of high-alumina concrete used in thousands of buildings in Great Britain. A mortar additive was used in some two thousand buildings in the United States in the 1970s to produce a structurally superior masonry wall. It is now known that this additive also corrodes the steel connections that tie the brick to the building structure. Similarly, certain additives used to preserve wood have caused fasteners to deteriorate.

Widespread problems have also resulted from unanticipated external effects. Parking ramps in northern climates have experienced severe structural problems, and many have had to be rebuilt. Salt used for melting ice and snow on the roads was brought onto the ramps by cars. It then migrated through the floors and corroded the structural steel reinforcing. The chemical PCB, commonly used in transformers as insulation, becomes a toxic gas when it burns, turning even a minor fire into a potential catastrophe. Large buildings, once contaminated, are unusable for years.

Unlike technical failures of building performance, ***functional*** *and* ***behavioral*** *problems are often* ***quiet failures*** *that may not be immediately evident.*

The functional and behavioral elements of building performance do not cause such high-profile problems, although occasionally even they produce outstanding revelations. Newman's well-publicized work (1973) clearly related building design and population attributes to safety and security in public housing.

Also controversial was the less than satisfactory performance of so-called open office and school design concepts, ideas that were enthusiastically adopted by architects and their clients in the 1970s. While these are now criticized for problems with privacy, distraction, and acoustics, there are also benefits to a more open environment that may be lost in the recent trend toward more traditional closed designs.

The Practice Model

Architects are being sued twice as often as they were in the 1970s, primarily for problems in building performance. The studies by Bonshor and Harrison (1982) and by Diekmann and Nelson (1985) indicated that a substantial number of problems may be due to design errors. Forty-three percent of architectural firms have lawsuits brought against them annually, and the average cost of settling a claim has risen substantially from $15,000 in 1975 to $40,000 by 1982. Many insurance companies no longer cover liability for architectural firms, and those that do have increased their premiums by as much as 300 percent. A significant number of firms can no longer afford to carry insurance.

Architects today are being ***sued twice*** *as frequently as they were ten years ago.* ***Most*** *of the suits have to do with* ***building performance*** *problems.*

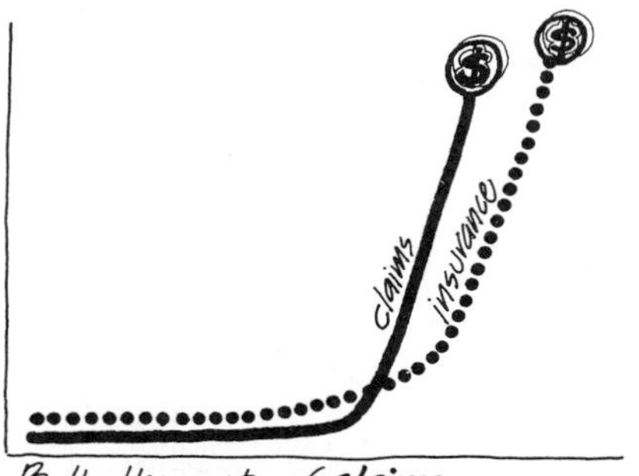

Both the costs of ***claims settlements*** *and professional* ***insurance*** *have risen* ***dramatically*** *in recent years.*

Traditionally, architects have defended themselves in ***lawsuits*** *by using the* ***precedent*** *of* ***"reasonable standard of care."***

The traditional defense of architects has been based on a "reasonable standard of care," in other words, "an average standard of skill and care of those of ordinary competence in the architectural profession" (Greenstreet and Greenstreet 1984, p. 9). Even if a performance failure due to faulty design occurs and is acknowledged, if the architect can prove that the work was carried out at a level reasonably expected of architects, he or she may not be held responsible. An architect does not guarantee satisfactory results. There is, however, a realization that the design and construction process is indeed very complicated and fragmented, lacking a reliable source of information on building performance and subject to special conditions such as location and the vagaries of weather. The architect is seen as designer and coordinator in a process in which there are many products and parties involved.

In recent years however, court rulings have been made using the "**implied warranty**" principle.

However, in recent years, courts have made rulings on the implied warranty principle. Here the architect is responsible for building performance and reliability, notwithstanding the standard of care of the average practitioner. A client is entitled to a building that works! Cases in some lower courts have found fault with architects based on the implied warranty standard, and although these have been overturned in the higher courts, they indicate a direction in rulings that may have significant implications for the profession in the future, such as raising the standard of care and broadening the areas of liability.

The increasingly aggressive legal environment for architects, the high insurance premiums, and increased exposure to liability (see appendix A) are only indications of some of the problems in building performance. The increase in lawsuits is only the tip of the iceberg; most cases are settled before they ever get to court.

Could such legal entanglements help produce better buildings in the future? Bernard Spring, head of the Boston Architectural Center, feels that

> The increase in litigation and insurance premiums is a free-market tool that really works. Architectural firms are now being careful in a way they never have been before (Spring 1987; correspondence with H. Z. Rabinowitz).

In our society, so much is built so quickly and so much change and innovation occur that problems in building design can be replicated many times before they are discovered. In some cases, decades can elapse before building problems are manifested, and ultimately many hundreds of buildings and their users can be affected.

Sometimes, problems in building design are **replicated many times** before they are **discovered**.

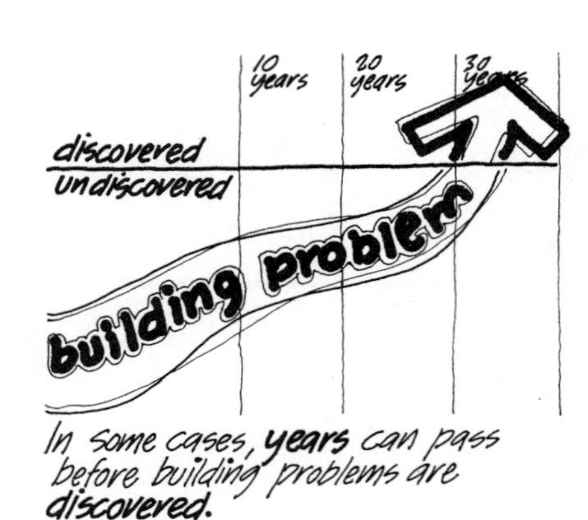

In some cases, **years** can pass before building problems are **discovered**.

A building may leak, its finishes may wear prematurely, and it may have problems with acoustics, lighting, and heat regulation. There may be poor air quality, sagging, delaminated or faded finishes, and the occupants may be too hot or too cold. A building may not be able to attract tenants or users, and it may have great difficulty being serviced or maintained. It may not have enough storage, or it may not be able to accommodate organizational changes.

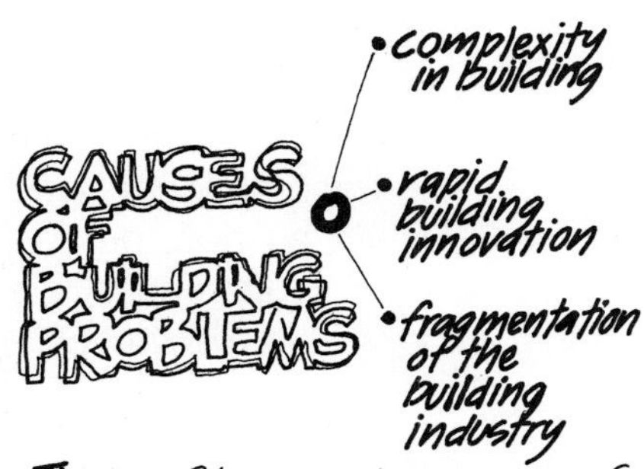

Three of the underlying causes of building problems are building complexity, building innovation, and building industry fragmentation.

It may not have adequate parking, or it may be difficult for people to find their way within it. Workers may be less productive because of a distracting environment or because of a building's layout. While conclusive studies have not been undertaken, enough evidence of serious, frequent, and pervasive problems in buildings has accumulated to be of major concern. What are the underlying causes of these problems?

COMPLEXITY IN BUILDING

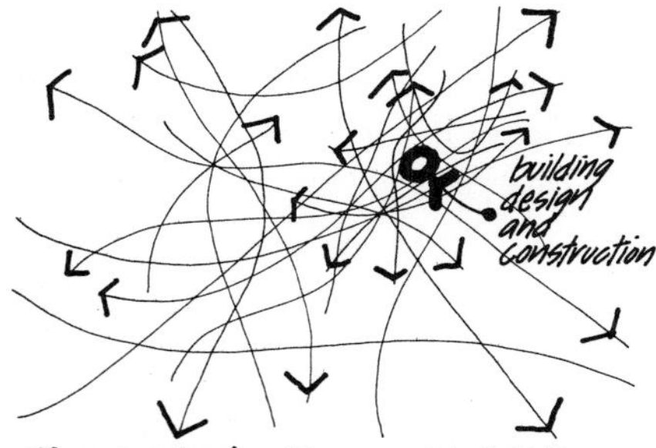

The context within which buildings are designed and constructed has become increasingly complex.

A review of the literature, an examination of the performance of materials and methods used in practice, and an overview of the legal environment for architectural practice provide evidence that the context in which buildings are designed and constructed can easily lead to buildings that are flawed in significant ways. In order to understand why this situation exists, it is critical to examine the context of the increasingly complex building industry.

> A process which involves so many people, connected by such a tenuous and opaque network of power relations and often working with such sophisticated and risky techniques, is bound to misfire from time to time (Prak 1984, p. 24).

As the sizes of individual buildings and groups of buildings increase, so do their costs. Projects in the $100 to $200 million category are not unusual, and the experienced professionals who previously would have been directly involved in the building process on a "hands-on" basis are now removed directly from the building scene. Prak (1984) summarizes this situation:

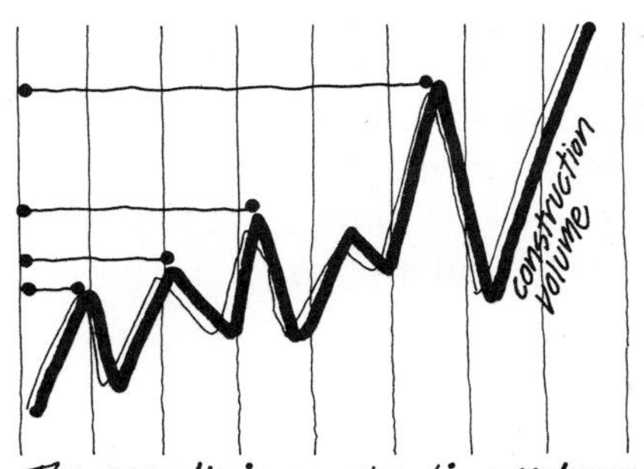

The growth in construction volume has put time and money pressures on clients, architects, developers, material suppliers, manufacturers, and builders.

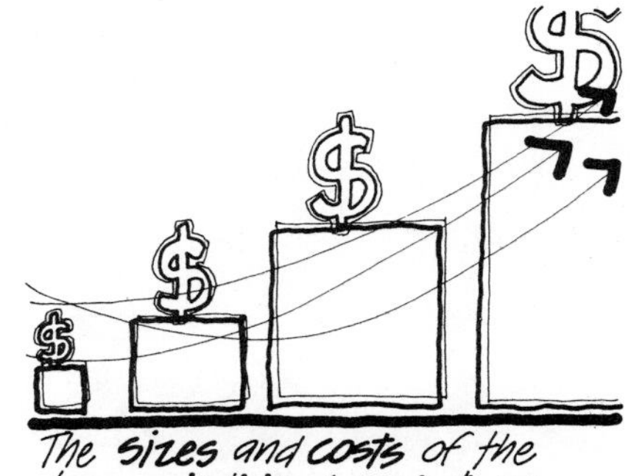

The sizes and costs of the larger individual projects are increasing.

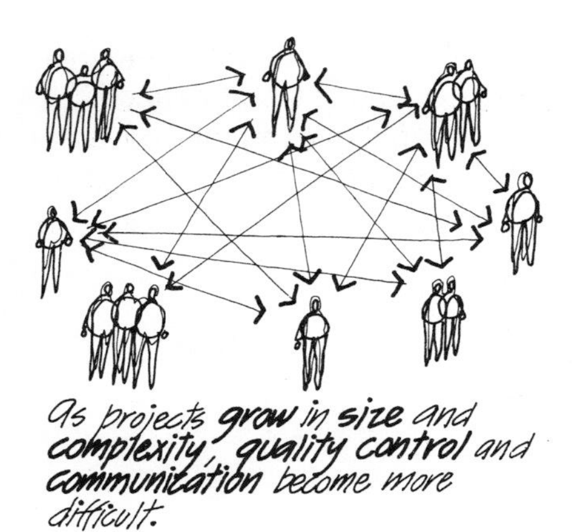

As projects grow in size and complexity, quality control and communication become more difficult.

Quality control was always a problem in the building industry. It grows worse when buildings increase in size, whenever larger crowds of journeymen are involved in the job, or when the pipelines between the main office, the job superintendent, and the crews become longer and longer. The organization for any building was always ad hoc, but the network of communications can wear so thin that it disappears. People no longer know what the other fellow is doing, and they often do not care to know (p. 23).

Innovation in the Building Industry

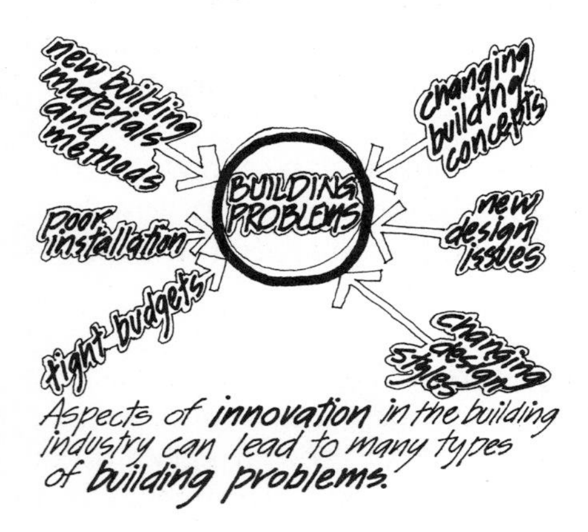

Aspects of innovation in the building industry can lead to many types of building problems.

New types of building materials and methods, as well as new types of building uses, are proliferating, and these, combined with poor installation and budgetary concerns, can often result in problems.

Building concepts also change over time. It is only since the mid-1960s that open office plans have become common. Now advanced space planning and "intelligent buildings," that is, buildings highly equipped for telecommunications and information processing, are being introduced. The form of housing continues to evolve as an emphasis on a diversity of lifestyles grows and the forms of ownership continue to change.

Some new concerns that add to this complexity tend to be generic, applying to all building types—energy conservation and accessibility for the handicapped, for example.

Design styles are subject to change. The public and architects tend to be increasingly interested in new forms and new building materials. Marble now frequently appears in architectural publications. Recently, stucco, multicolored brickwork, metal roofs, classicist design motifs, pediments, exposed trusses, neon lights, pergolas, unfinished concrete block, exposed chain-link fencing, brass, and granite have become quite popular. An interest in history has stimulated the use of eclectic forms and materials. The forms present in current stylistic expression range from primitivist huts to canted walls inspired by Egyptian temples to neon lighting from a 1950s diner. These stylistic changes have occurred on an almost continuous basis in recent years so that adequate learning of appropriate construction techniques may not have occurred before one style is supplanted by another.

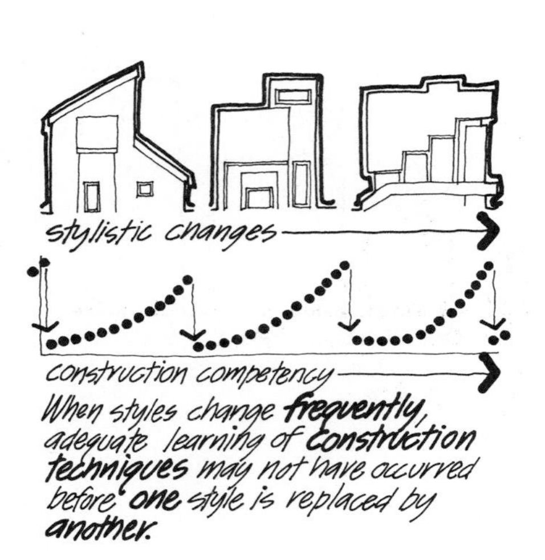

When styles change frequently, adequate learning of construction techniques may not have occurred before one style is replaced by another.

Fragmentation of the Building Industry

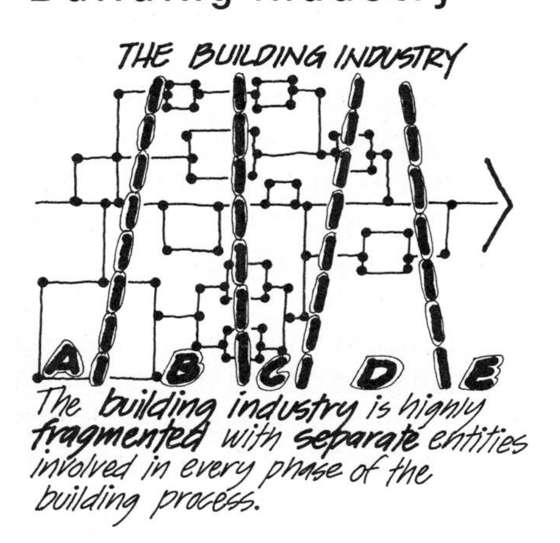

The building industry is highly fragmented with separate entities involved in every phase of the building process.

Changes in volume, project size, materials, concepts, and design are exacerbated by the highly fragmented nature of the building industry, in which separate entities are involved in every phase of the building process, from land acquisition, development, design, engineering, and construction to leasing and sales, building operation, maintenance, and management. In addition to vertical fragmentation, there are many horizontal divisions among materials suppliers, labor organizations, and various consultants. In a large project, the design aspects alone involve the coordination of many different engineering firms (civil, electrical, mechanical), as well as illumination, acoustics, landscaping, parking, leasing, and interior design consultants. Ordinary communication among this large and assorted cast of characters is a very difficult undertaking. Again, Prak (1984) comments:

> Many threads come together in the architect's hands: form, construction, cost, and supervision. But the position looks stronger than it really is: the organization of the building team is weak and temporary. There is a similarity here to the influence of architecture on the behavior of the user: a real influence exists, but it also looks more than it really is (p. 32).

RATIONALE FOR POE

Facilities Management

Buildings are major resources that have often been little understood and inadequately managed. However, a new field of facilities management has developed to integrate the design, economic, and functional aspects of buildings. Buildings are resources in two ways: they are a direct economic resource constituting about 25 percent of corporate assets and investments, about 33 percent of personal wealth, and about 50 percent of all tangible wealth in the United States. Furthermore, the form and construction of buildings affect the efficiency, productivity, and enhancement of the lives of their occupants, and as such buildings are critical resources.

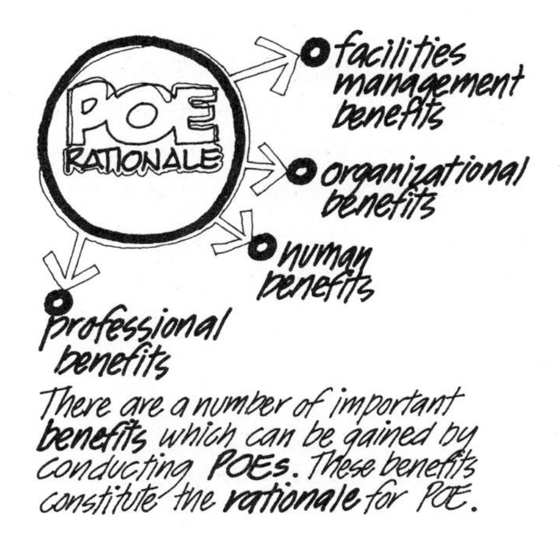

There are a number of important benefits which can be gained by conducting POEs. These benefits constitute the rationale for POE.

POE can help facility managers to manage better the existing building stock of their companies.

Much of this chapter has focused on how well buildings work, yet buildings as a resource have been given little formal attention. In the past, characteristics of the building industry, among other factors, have prevented the proper management that facilities deserve as economic assets. Although there have been no studies of building demographics on a suitably large scale, there is mounting evidence that significant improvements can be made regarding the technical, functional, and behavioral elements of building performance.

Only 1 to 2 percent of our building stock is replaced annually, which would seem to make evaluation and feedback only marginally effective. However, the growing importance of renovation and adaptive reuse, as well as the new attitude toward managing real estate assets, has changed this situation. Previously, building owners and developers have been small firms that were local and regional in scope. Today, large-scale developments are often carried out by a growing number of very sophisticated and substantial firms that are involved in long-term projects and operate on a national scale. They are responsive to markets and manage most of their own assets. These changes in the building industry have been important in the growing acceptance of POE.

Processes such as POE are successful when they can demonstrate positive results. Such benefits are easily demonstrated in the problem-laden building industry. For instance, the construction management sector did not exist in 1967, but today its dollar volume rivals that of architectural firms. Construction management improved coordination in the fragmented building industry. Similarly, the increased use of POEs in institutions is beginning to be seen where the benefits of POEs are shared by the organizations that sponsor them, the occupants of the buildings being evaluated, and design professionals.

Organizational Benefits

*POE can **help** organizations **test** new building ideas and **operate** more **efficiently** within their facilities.*

The benefits of POE to organizations are becoming clearer as the field has evolved from academic to applied research and as economic measures of efficiency and productivity have been developed. Early work conducted at the Building Performance Research Unit in Scotland (Thorne 1980) and subsequently at the National Bureau of Standards noted that building design and construction constitute a rather small percentage of total operational costs. For example, in office buildings, personnel costs are over 90 percent of organizational expenses over time, while the original construction costs are in the 4 to 5 percent range. The leverage involved is considerable: small increments in building costs may result in large savings or improvements in organizational effectiveness. These percentages vary for different functions. In schools, for instance, percentages are similar to office buildings, while in manufacturing plants and hospitals there is less leverage.

Other benefits are more directly building-related. The needed reduction in technical problems is obvious. In addition, POEs in office buildings and hospitals provide a strong economic rationale for including flexibility in certain building functions. The Marriott Corporation conducted a comprehensive POE of a new prototype hotel that was field-tested on hotel guests in order to determine optimum room size, choose appropriate amenities, and minimize operational expenses. As a result, hundreds of these "courtyard" hotels will be built in the United States over the next decade (*Wall Street Journal* 1985, p. 1). Given better understanding of the benefits of the POE process, large-scale organizations such as supermarket chains, franchisers, and developers of offices, retail centers, and housing are increasingly employing POE methods. As a result of POEs, the improved design criteria and guidelines will lead to improved design decisions and thus will contribute to better quality buildings. Market forces will be a major reason for the use of POEs in the future.

Human Benefits

POE can benefit building **occupants** by helping to make environments **humane, appropriate, obstacle-free, enjoyable, safe,** and **responsive.**

Perhaps the most important benefit of POEs is their positive influence on the creation of humane and appropriate environments for people. Individuals can benefit from the use of POEs in many ways: through the avoidance of problems that are obstacles to the effectiveness and enjoyment of workplaces and living environments; through the development of an appropriate design response that helps support activities; through careful performance specification, ranging from human factors in doorknob designs to sensitive planning of major multiuse complexes; through the generation of an appropriate social climate; and through the creation of appropriate image and meaning in the built environment that may enhance the lives of its occupants.

Professional Benefits

POE can benefit **professionals** by reducing **legal** problems, **increasing** the **excellence** of their **products**, and helping them make **informed** decisions.

POE allows professionals to make informed decisions about the design of the built environment. The systematic and comprehensive evaluation of buildings, the development of credible performance criteria, and the availability of design guidelines will lead directly to significant improvements in the quality of buildings. The widespread use of POEs will also help reduce the legal and liability problems of the profession.

The problem of building performance looms large in spite of, and sometimes because of, well-intentioned design professionals. Architectural practice and maxims—a closeness to the client, a bias toward action, a philosophy of excellence—represent the skills and ideals required to carry out POEs and to implement their recommendations. The systematic development, dissemination, and use of programming and design information (Building Research Board, 1986) will provide the benefits of professional excellence needed to match the level of professional idealism.

However, POE does not necessarily guarantee proper building performance: the building industry will still be fragmented and the creative skills of the designer are still critical.

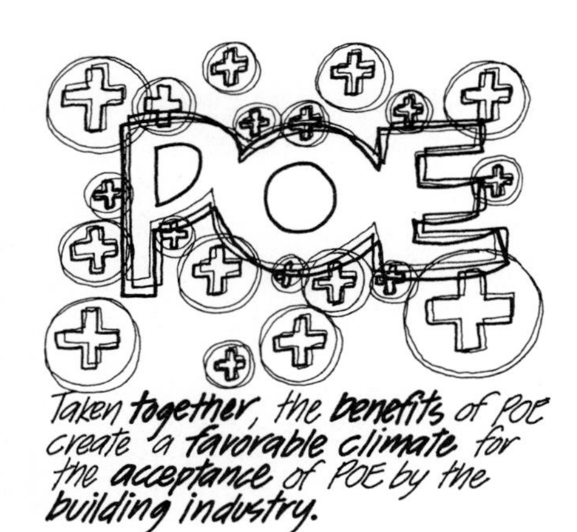

Taken ***together****, the* ***benefits*** *of POE create a* ***favorable climate*** *for the* ***acceptance*** *of POE by the* ***building industry.***

In summary, the climate for the acceptance of POEs by participants in the bulding industry is now excellent for several pragmatic reasons: increasing problems in building performance; problems in legal and professional areas; changes in the development and building industries; the recognition of buildings as important strategic assets that need to be managed; and the multiple benefits that are to be derived from well-conceived and user-responsive buildings (Building Research Board, 1987). POE has developed since the 1960s to the level where there is a critical mass of both knowledge and practitioners (Kantrowitz et al. 1986), as well as an appreciation of the technical, functional, and behavioral improvements that POEs bring to building performance.

3. The Building Performance Concept

The concept of building performance is the major philosophical and theoretical foundation of POE. It is no coincidence that the professions of construction management and facilities management, both concerned with specific building performance, emerged parallel with efforts in the areas of building evaluation and environment and human behavior. The earliest applied work

The philosophical and theoretical foundation of POE is the concept of building performance.

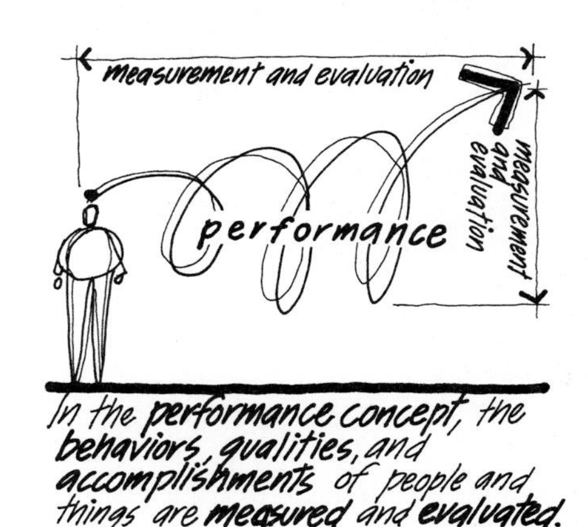

In the performance concept, the behaviors, qualities, and accomplishments of people and things are measured and evaluated.

on building performance in the United States was conducted by Ezra Ehrenkrantz and his associates on the School Construction Systems Development Project in California (Educational Facilities Laboratories 1967). This work was further advanced at the Institute of Advanced Technology at the National Bureau of Standards. There, under the direction of John Eberhard, the "performance concept" was generated (Eberhard 1965). Subsequent projects (Wright 1971) executed by the National Bureau of Standards for the Department of Housing and Urban Development and the General Services Administration built upon these initial efforts.

What is the performance concept and why has it made such an impact on the building industry? The peformance concept is used, for instance, when comparing the hitting performances of baseball players by using their batting averages. There may be other subtleties involved, such as power hitting or including the number of intentional walks a player receives, or comparing performance today with that of baseball stars of the past. A horseracing handicap or tip-sheet has a great deal of measured data about a horse's past

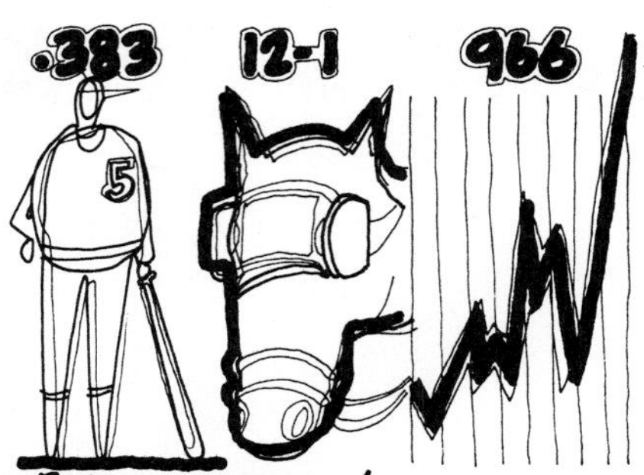

Batting averages, horse racing odds, and stock market averages are all examples of the performance concept.

performance, enabling the bettor to compare criteria, make an evaluation, and place a more informed bet. The stock market is also performance-oriented. Multiple performance measures are used, and analysts regularly compare their performance measures with such criteria as the Dow-Jones average.

While both the idea and the application of the performance concept are quite common, in general this concept was only recently accepted for widespread use in the building industry (Leslie 1985; Building Research Board 1987).

The performance concept has only recently been applied to the building industry.

How were building design decisions made previously? To use the racetrack analogy again, sometimes decisions are based on such whimsical criteria as the color of the jockey's silks rather than on measures of past performance. Comprehensive, satisfactory building performance information was largely unavailable or unused. Windows, for example, were chosen on the basis of tradition, price, or an appealing brochure. The windows' performance in terms of insulative qualities, infiltration rates, cold bridges, weatherability, cleanability, durability, and other factors was not addressed. As a consequence, a prescriptive, not a performance, basis was used in design decision making.

Evaluative procedures are, in large part, the bases for professionalism and credibility in law, medicine, and business.

In contrast, the fields of law, medicine, and business use evaluative procedures that are in large part the bases for their professionalism and credibility. Law, for example, has a systematic, codified, and formal approach to evaluation through the use of precedents or measures of past performance in determining legal arguments and decisions. The practice of medicine has its basis in continuing research, quality assurance procedures, and publications. Given a patient's medical history, age, sex, weight, and other attributes, a physician can provide a disease evaluation and prognosis. Substantial evaluation information and criteria are available and routinely used in every subspecialty of medicine. Business education stresses evaluation through the use of the case-study approach as well as simulation models. All three fields—law, medicine, business—have made substantial progress and achieved a high degree of professionalism because of their rigorous use of evaluation and feedback in assessing past successes and failures in their respective areas of specialization.

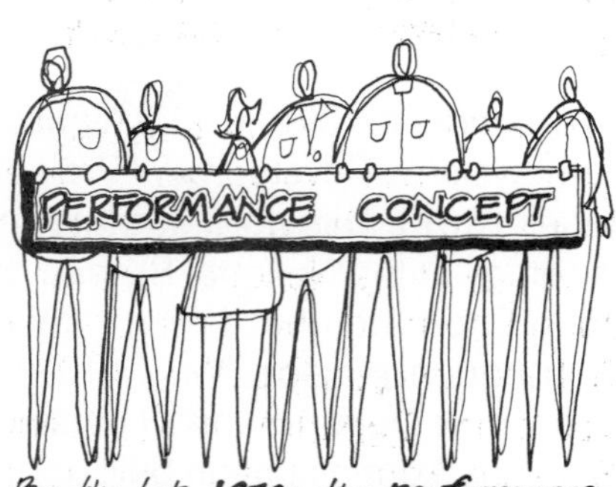

By the late 1970s, the performance concept was used throughout the architecture profession.

By the late 1970s, the performance concept was used throughout the architectural profession. The energy crisis of the mid-1970s helped accelerate this change. New research results were also translated into performance-based evaluation criteria, for example, in the measures to prevent flame spread and to contain smoke developed in fires. Architects had used some performance measures in the past, such as fire resistance or sound transmission, but their application was rather limited. Now, performance measures are used much more comprehensively (Brill 1984).

ADVANTAGES OF MEASURING PERFORMANCE

The performance concept, as depicted in figure 3-1, has the following characteristics: in the act of evaluation, performance measures are compared with appropriate performance criteria, and a conclusion is reached on how successful the building performance has been. An evaluation, combined with recommendations for improvement, is used for feedback and feedforward regarding the performance of similar building types.

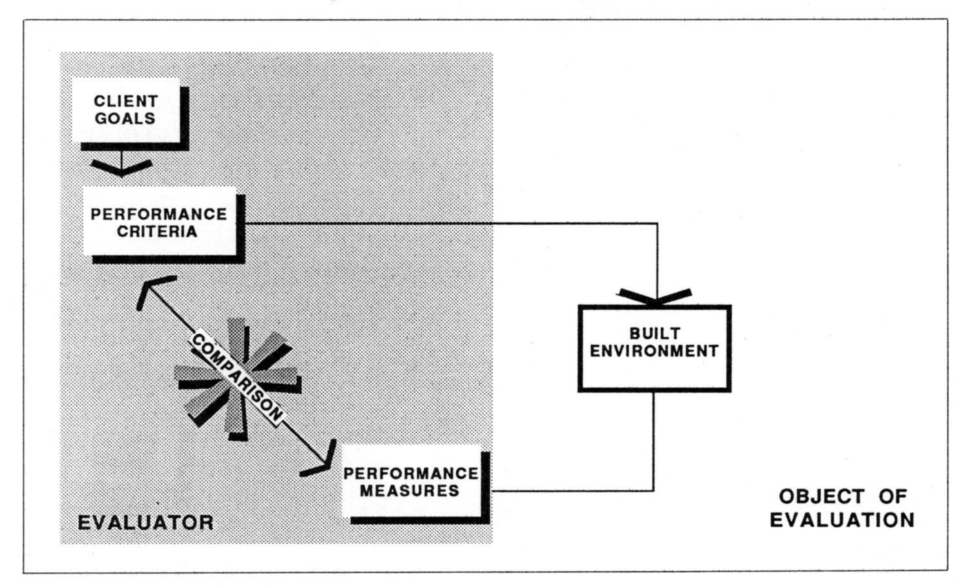

Figure 3-1. The performance concept.

Performance evaluation and feedback relate client goals and performance criteria to the objectively and subjectively measurable outcomes of buildings (Preiser 1986). In figure 3-1, everything shown in the shaded area is dependent upon the *relativity* of person/environment relationships. The same building and its physical attributes that can be objectively measured and described may be perceived by the same people differently at different times, or differently by different people at the same time. Cultural relativity affects this, and therefore, all one can realistically deal with in POE or in the programming and design of buildings is the perceived performance of the building.

A second and important feature is that the evaluator is inside the shaded area of the figure, implying again the relativity of perceived building performance. The evaluator is the driving force of the evaluation system and thus introduces biases, sets the scope of the evaluation, and presents the findings to the client. These qualifications indicate that there are no absolutes in environmental design.

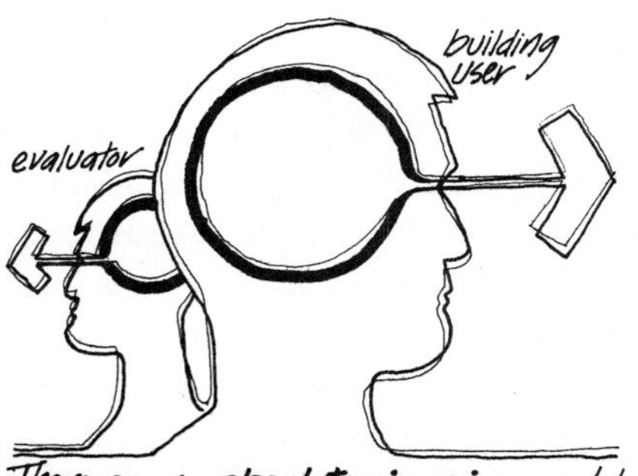

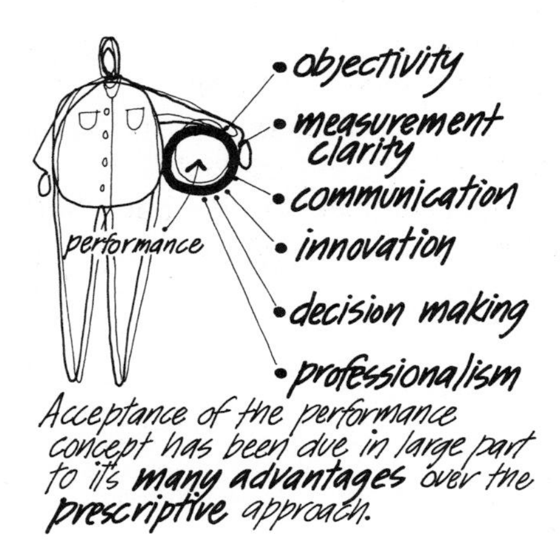

Acceptance of the performance concept has been due in large part to its many advantages over the prescriptive approach.

Notwithstanding these qualifications the performance concept has been accepted because of its many benefits:

- *Increased objectivity:* The performance concept engenders objectivity because opinion is replaced by measures of performance (ASTM 1986).
- *Clarity of measurement:* Measured building-performance information and criteria clarify what factors are relevant in design decision-making.
- *Enhanced communication:* Relevant criteria and measures of performance can be understood and discussed by the many participants in building because of standard measures, language, and tests.
- *Incentives for innovation and the development of alternatives:* The use of performance criteria allows the development of a range of solutions to design problems as long as the solutions meet the relevant performance required.
- *Aid in decision making:* A rigorous and objective analysis can be made of the relative merits of alternatives at a detailed level.
- *Advanced professionalism:* Dissemination and use of performance-based information, the expansion of performance information into new areas of knowledge, and the evaluation and refinement of performance measures and criteria all contribute to professionalism in the building industry.

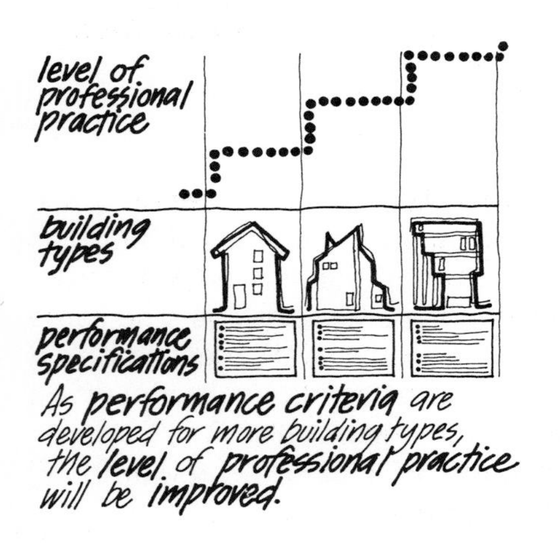

As performance criteria are developed for more building types, the level of professional practice will be improved.

These advantages are significant to the building industry and the architectural profession. Performance-based products, assemblies, methods, and configurations aid the architect in generating building alternatives and design iterations. As performance-based measures are used and criteria developed for more building types, the level of architectural practice will be improved.

The results of using the performance concept in building depend on the client, the evaluator, the nature of the performance criteria, and their effects. For example, widespread building-performance problems or solutions affecting safety or health will have appropriate performance-based regulations included in the four major regional building codes and in local and state codes.

Moreover, a specific building problem concerning functional performance, such as the level of generated static that can affect computers, may be examined by the manufacturers and users of computers as well as building-product manufacturers. There is now an effort to develop a standardized method to measure static buildup, and a maximum level of about 2 kilovolts is presently seen as acceptable. While the criteria and test methods do not have the force of legal regulations, it is in the best interest of any professional involved in designing facilities that will use computers to choose products and methods that meet these industry-set functional criteria.

A third advantage, and a less formal application of the performance concept, would be in a specific project done by an owner or designer. "Benchmark" criteria from similar projects or exemplars may be used. Such criteria may have neither the force of legal standards nor of those published by industrial groups, but they have great influence in the marketplace. For instance, in the design of an office building, the owner may set criteria of 87 percent leasable area and a 45,000 annual Btu-per-square-foot energy load based on benchmark competitive buildings. In a retail facility, the developer and architect may be concerned with performance in deciding which amenities would best attract shoppers. Criteria concerning functional distance may be a measure for a facility where social interaction is important. Such criteria are now part of a "commercial vernacular" passed on by word of mouth or discussed in trade magazines. These norms may be formally untested. Despite that fact, these criteria are useful and recognized as valid, though short-range, measures of performance.

Performance Evaluation

The term *evaluation* contains a form of the word *value,* which is critical in the context of POE since an evaluation has to state explicitly which and whose values are used in establishing evaluation criteria. A meaningful evaluation focuses on the values behind the goals and objectives of clients or those who carry out the evaluation. Thus, performance criteria used in evaluation are developed from goals and objectives that in themselves are derived from values held by individuals, groups, organizations, or entire socio-political

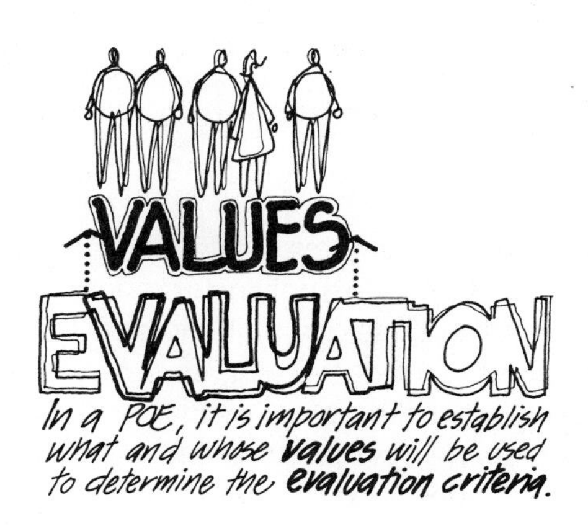

In a POE, it is important to establish what and whose **values** will be used to determine the **evaluation criteria**.

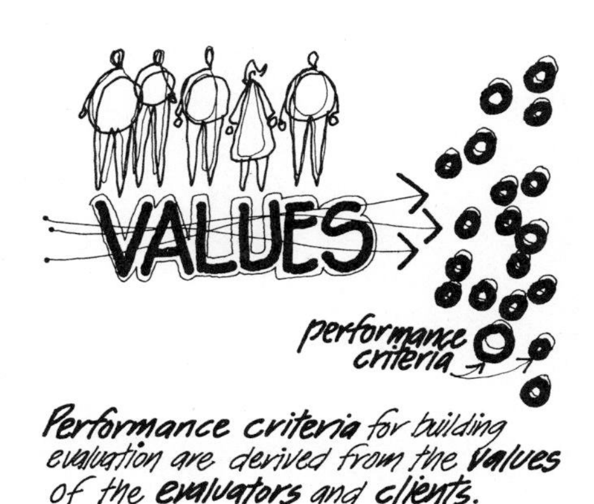

Performance criteria for building evaluation are derived from the **values** of the **evaluators** and **clients**.

systems. Frequently, there are differences among various groups or organizational units. Resolving such conflicts helps in achieving satisfactory building performance. The establishment of a clear linkage between values and POE also implies an enlightened, proactive stance for those who manage facilities with a desire to improve the building stock and general understanding of building performance.

There is a difference between the quantitative and qualitative aspects of building performance and the respective performance measures. Many aspects of building performance are in fact quantifiable, such as lighting, acoustics, temperature and humidity, durability of materials, and the amount and distribution of space. The qualitative aspects of buildings, such as esthetic beauty or visual compatibility with a building's surroundings, are more difficult to evaluate. In this case, the expert, evaluator, or connoisseur will pass judgment. Examples are the expert ratings of scenic and architectural beauty awarded châteaux along the Loire River in France, as listed in the travel guide

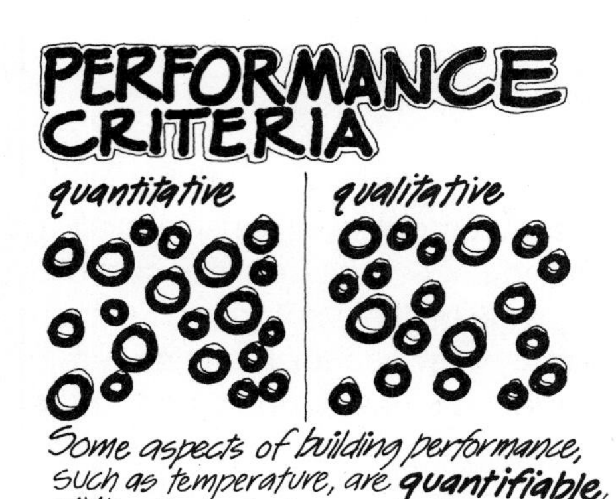

Some aspects of building performance, such as temperature, are **quantifiable**, while others, such as esthetics, are **qualitative**.

literature. The higher the apparent architectural quality and interest of a building, the more stars it will receive. Consequently, one can expect that those ratings will be reflected in the numbers of visitors who flock to such tourist attractions. There are similar expert ratings of the quality of restaurants, also affecting the number of customers attracted and therefore the size of their revenues. Recent advances in the assessment methodology for visual esthetic quality or scenic attractiveness provide hope that someday even this elusive domain will be treated in a more objective and quantifiable manner (Nasar 1988).

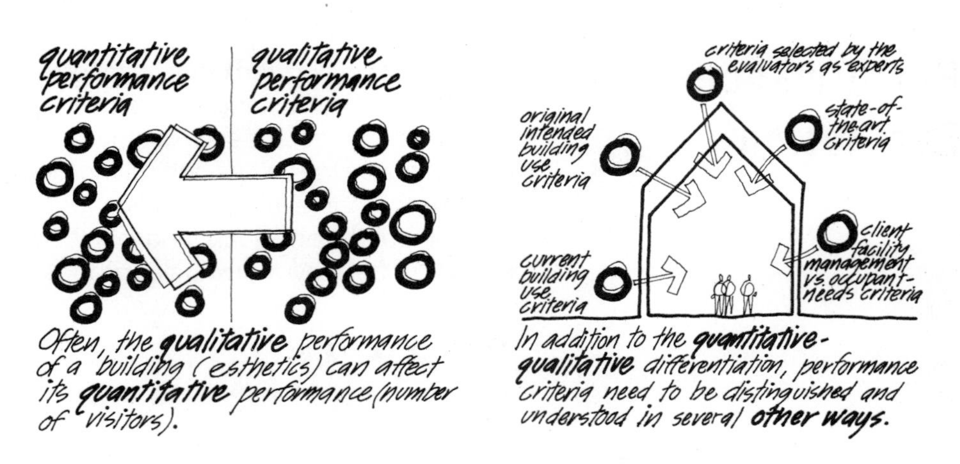

Often, the **qualitative** performance of a building (esthetics) can affect its **quantitative** performance (number of visitors).

In addition to the **quantitative-qualitative** differentiation, performance criteria need to be distinguished and understood in several **other ways.**

The need for explicitly stated performance criteria was emphasized by Davis (1970). In applying criteria, one needs to determine the criteria for both the current use of a building and its originally intended use. Also needed are the criteria that the evaluator or evaluator team may apply as experts on a certain building type, or that relate to the client organization's management versus the occupants' criteria for the desired performance of a facility.

The need for evaluation criteria should not obviate some of the realities that do occur in the course of POEs. For example, the evaluator team may deal with a new facility type for which no evaluation criteria exist, or it may encounter unexpected phenomena that turn out to be critical for the operation of the facility. How is evaluation accomplished? An analogy to the design process is applicable, that is, the need to close an information gap creatively or to devise solutions for which inadequate information such as design criteria exist.

THE BUILDING PROCESS AND THE PERFORMANCE CONCEPT

Applying the performance concept to the building process means using the principles of the performance concept—measurement, comparison, evaluation, and feedback—in designing and building new structures. It is part of a systematic approach to improve the quality of the built environment. Furthermore, it is a movement that includes a variety of mechanisms intended to make buildings more responsive to the functions they support and to the needs of the buildings' occupants.

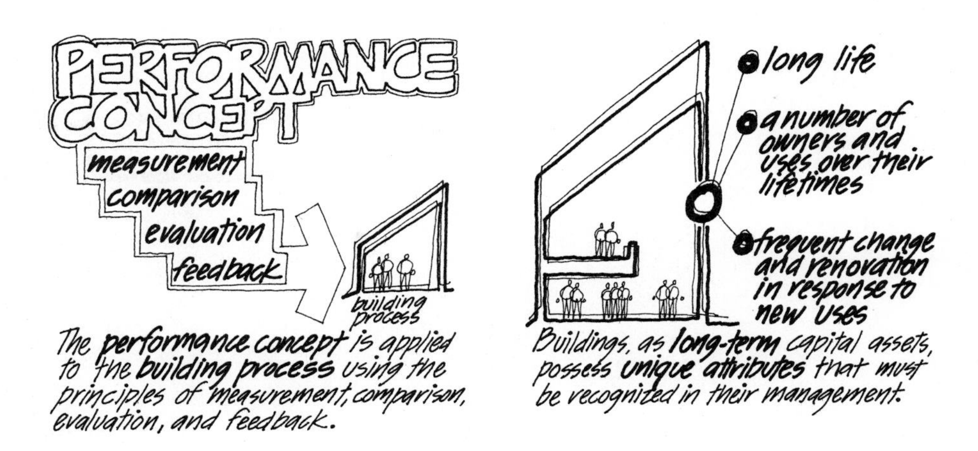

*The **performance concept** is applied to the **building process** using the principles of measurement, comparison, evaluation, and feedback.*

*Buildings, as **long-term** capital assets, possess **unique attributes** that must be recognized in their management.*

Buildings have attributes that are not typical of capital equipment such as cars or machinery. The managers of facilities, in their effort to create and operate environments that will produce high-quality performance and satisfy their occupants, must respond to these attributes:

- Buildings have a much longer life than most assets in business. A useful life of forty years is considered minimal for most buildings.
- Buildings involve a number of owners, organizations, and users throughout their lives. On the average, a home is owned by a single owner for about eight years, and the average tenancy in apartments is about three years. Ten or more organizations may occupy the same area in an office building over its life, while an office worker may be relocated every two to three years within his or her organization.
- Existing buildings are being changed and renovated more often in response to new owners, organizational changes, and new occupant requirements.

*The **performance concept** is **central** to the aggressive and proactive **management** of building assets.*

Chapter 2 discussed the emerging trend in the aggressive management of building assets. This type of management affects all aspects of the building process from planning and programming to occupancy throughout the building's life cycle. The performance concept in the building process views buildings as dynamic entities and indicates a comprehensive attitude toward the management of buildings.

Figure 3-2 contains the same elements as the basic performance concept shown in figure 3-1. Performance is measured, compared to criteria, and the evaluation results are used as feedback to improve the evaluated building's performance.

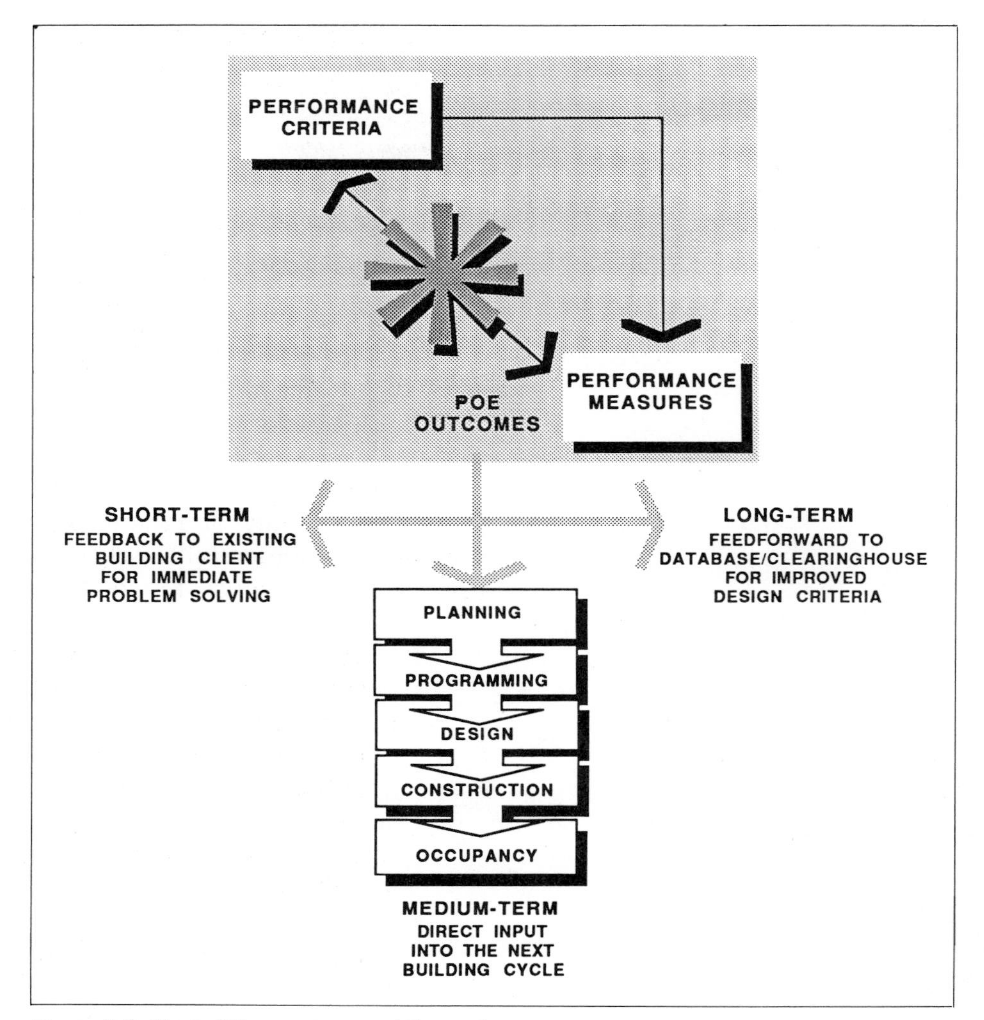

Figure 3-2. The building process and the performance concept.

However, the diagram goes beyond simple feedback. Feedforward is also critical. The planning, programming, design, and construction of future buildings can be improved through the feedforward of POE results. The creation of databases or information clearinghouses on building types, attributes, and occupant groups can benefit building owners, users' organizations, and design professionals alike in order to improve both new and existing buildings.

The feedback and feedforward functions in the performance evaluation research framework described below direct the results of POE to the various phases of the building process. For instance, an acoustical evaluation can be relevant to the programming of a building (setting the basic acoustical criteria), design aspects (location of rooms, detailing, and selection of materials), construction (installation), and occupancy (furnishings), as well as during the life of the building (maintenance of acoustical attributes). A problem in this area can be attributed to almost any phase in the building process, and an important part of POE is to determine which of these phases need improvement.

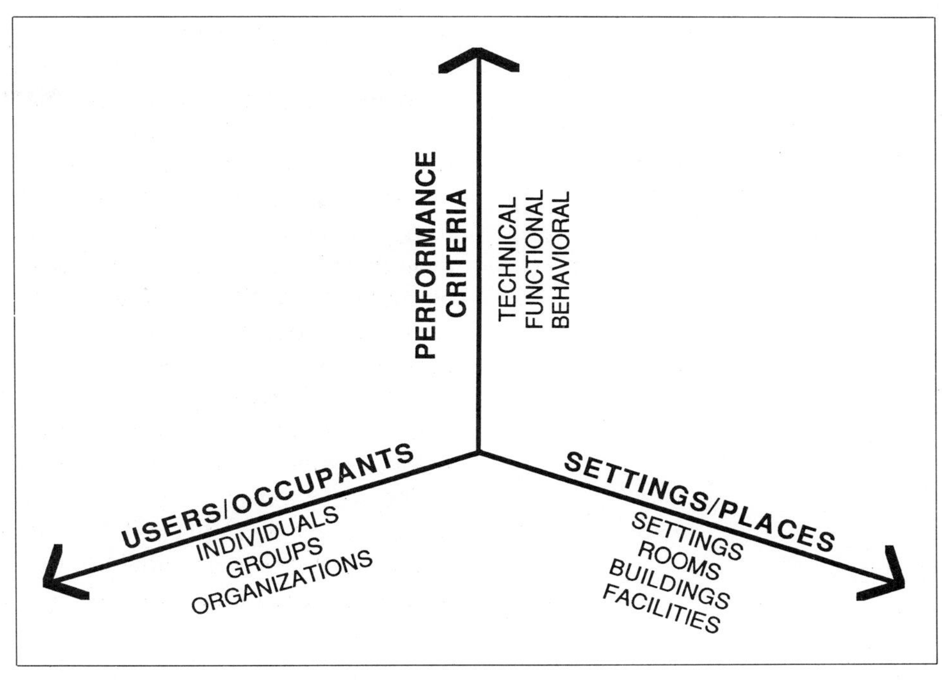

Figure 3-3. Elements of building performance.

ELEMENTS OF BUILDING PERFORMANCE

The third aspect of POE in the building process pertains to the elements of performance that are measured, evaluated, and used to improve buildings (fig. 3-3).

The three major categories of elements in building evaluation relate to technical, functional, and behavioral performance. While there are other factors such as location and economics, the above three elements are the most important in terms of physical performance implications affecting owners, organizations, and building occupants.

One of the horizontal axes in figure 3-3 indicates the building scale at which the performance in the three categories is measured and evaluated. In ascending order of scale, there are specific settings, rooms, buildings, and facilities. Acoustical performance, for instance, is measured in the building as a system (transmission between rooms), within rooms (ambient sound), and in particular settings (acoustical distraction due to unwanted sound within an open space). Circulation would primarily be measured at the level of the system as a whole. The adequacy of display space in buildings can be measured in each or all of the above three categories. Seating would normally be assessed on a setting-by-setting basis and not throughout the building.

The elements of performance comprise the basic attributes of buildings. Although the elements of performance in this framework consist of discrete evaluation criteria, based on objective tests and measures, in reality many elements act together to create the overall environment as experienced.

The second horizontal axis in figure 3-3 differentiates among users/occupants of buildings in terms of their numbers or other key characteristics such as age group, sex, lifestyle, race, or status within an organizational structure, to name just a few. Another important aspect is whether users/occupants of buildings are individuals, groups, or entire organizations.

The vertical axis in figure 3-3 contains performance criteria for the three categories of elements that are considered in buildings. The most important ones are described below, broken down into the categories of technical, functional, and behavioral elements of building performance.

Technical Elements of Building Performance

Technical elements can be characterized as the background environment, a kind of "stage set" for activities. These elements include the survival attributes—structure, sanitation, fire safety, and ventilation. They deal with health, safety, and welfare and often appear in building codes. Additional elements include the performance of roofs, walls, finishes, illumination, acoustics, and environmental control systems, including heating and ventilation.

Technical elements of building performance are often unnoticed because they form the background or stage set for the occupants' activities.

There are many evaluation procedures for single building elements and components but far fewer procedures for assemblies of multiple components.

Technical elements can be measured by instrumentation, and there is a considerable infrastructure of organizations and information to help accomplish this task. Such evaluation has occurred for some time, with measures, performance criteria, and in-place evaluation processes for specific building materials. However, there are few processes for the evaluation of assemblies of building materials, in a wall for example, or ensembles of larger components, such as in a specific activity area or space category (workstation, classroom, and so on). Another drawback of existing procedures is that most testing and evaluation take place in laboratory settings rather than in the field, and it is frequently the interaction of materials under actual-use conditions that leads to building problems.

FIRE SAFETY

Fire safety in structures was probably the earliest element to be evaluated systematically. The results of such evaluations are actively used because of concern for life and property. Relevant criteria include the fire resistance of the major structural elements of a building, fire extinguishment and containment, flame spread, smoke generation, the toxicity of burning materials, and the ease of egress in case of a fire.

This element of performance is well researched and documented, with much information based on actual fire experience. However, criteria are still being developed in a number of areas: the human aspects of finding one's way and exiting under fire conditions; the specific needs of such user groups as the elderly; how to measure fire toxicity; and the "fire load" (the potential fuel) generated by different user groups. For instance, an architectural office may have a large amount of exposed paper that can be dangerous under some conditions and may need specific precautions for fire safety.

STRUCTURE

This element, because of a considerable history of research and its presence in codes, has very high reliability. Structural properties, strength of materials, connections, and the durability attributes of structures are all contained in building codes. Evaluation occurs on-site as well as in laboratories, and feedback is pursued. Recent improvements regarding this performance element are primarily in advanced research areas such as wind effects on tall buildings, seismic design, long-span structures, and the interactions of materials that can affect structural connections.

The tragic failure of the skywalk at the Kansas City Hyatt Regency Hotel has also stimulated the examination of the human element in structural design and construction. The efforts of many parties involved in achieving safe structural building performance must be coordinated. An emerging concept is peer review in engineering, to provide quality control and to assure an appropriate level of performance in complex projects.

SANITATION AND VENTILATION

The elements of plumbing and ventilation represent highly reliable systems that have well-developed performance criteria. However, lowered ventilation criteria due to the energy crisis, combined with tighter buildings, have made indoor air quality an important area for POE research, as described in chapter 2. A very recent concern is the level of radon exacerbated by low rates of ventilation in the indoor environment. Radon is an odorless gas resulting from the decay of minute levels of uranium in specific geological areas. Higher-than-acceptable levels of radon are being found in thousands of homes as more research is done in this area.

ELECTRICAL

Electrical systems are a highly reliable performance element regulated by building codes and subject to technical changes and new functional needs, especially in office buildings. These changes include new equipment, such as microcomputers, facsimile communications, and the like, but also new technologies in cabling, such as flat wiring and fiber optics.

EXTERIOR WALLS

Exterior walls are almost always an assembly of materials. They include numerous joints and are exposed to elements that tend to cause them to deteriorate. Aside from energy, structural, and fire-related considerations, much of wall performance is unregulated, although many product manufacturers' organizations have developed performance criteria that include weathering, fading, moisture and wind infiltration, spalling, buckling, delamination, cracking, cleanability, and erosion.

ROOFS

Roofs represent the most frequent building failures. They are the most exposed of all building elements and, like the exterior walls, are an assembly of various products.

Despite the development of performance criteria and specifications, as well as frequent seminars for those concerned with roofing design, construction, and maintenance, problems still occur. New types of roofing, such as single-membrane synthetic products, have recently been introduced that may eventually alleviate the endemic problems.

INTERIOR FINISHES

The performance criteria for interior finishes (the floor, ceiling, and walls) emphasize such aesthetic attributes as fading, evenness of surfaces, and cleanability. Durability, particularly of floor surfaces, refers to resistance to scratches, indentations, abrasions, spills, stains, cigarette burns, and other damage. Factors such as replaceability, resistance to vandalism, changeability, and resilience may also be important in certain situations.

The three interior finishes act as an ensemble to modify substantially the acoustical and lighting environment. Thus, there are criteria for interior finishes regarding absorption and reflection of both light and sound.

ACOUSTICS

Acoustical considerations are infrequently found in building codes, although there are federal regulations governing sound in the workplace and the protection of workers exposed to noisy environments. Acoustic criteria cover the ambient level of sound, the transmission of sound between areas and rooms, reverberation, and specific areas such as machine noise and auditorium acoustics.

Criteria in acoustical performance are changing—particularly for offices. The open office is being reconfigured based on recent POEs, and solutions are being developed combining the benefits of communications between employees found in the open-plan office with an acoustical performance that allows for concentration and productivity.

ILLUMINATION

Standard illumination practice goes far beyond the minimal levels specified in building codes. Criteria exist for the quality and quantity of illumination based on human comfort and high levels of visual acuity.

More advanced criteria are now being developed. Physiological effects of the color of light, types of bulbs, increased use of video displays, need to change focus, and increased user control of task lighting are being studied to improve existing criteria.

ENVIRONMENTAL CONTROL SYSTEMS

Basic criteria for thermal comfort have been in existence for years and have been extensively tested. The energy crisis of the early 1970s changed many of the criteria, and new approaches to comfort, new technologies, and new variables were added in this area. As discussed in chapter 2, many changes were initiated and implemented without knowledge of their potential side effects.

Functional Elements of Building Performance

Buildings house organizations, such as businesses, factories, schools, fire departments, shops, or offices. These organizations expect satisfactory functional performance from their buildings. Functional elements include access for personnel and equipment, security, parking, and adequate spatial capacity for the activities to be accommodated by the organization. Other elements include utilities, telecommunications, adaptability to accommodate necessary equipment, responsiveness to change over time, and efficiency of communication and circulation.

The functional elements of a building directly support the activities within it, and they must be responsive to the specific needs of the organization and occupants, both quantitatively and qualitatively. Structural criteria, for instance,

Functional elements of building performance have to do with the **organizations** and **activities** that buildings contain.

Client **organizations** and **activities** should be **supported** by the performance of a wide range of **functional** building attributes and elements.

may not change much whether the building is a supermarket or a school. However, functional criteria such as storage will be completely different.

Although many elements of functional performance are generic such as the above-mentioned storage, necessary to most organizations, some elements are highly specific to building types. In a hospital, for example, specialized functional elements include containing infectious water, shielding radioactive areas, distributing oxygen for the movement of nonambulatory patients, and transporting meals from a kitchen area to patients' rooms.

Some facility types, such as hospitals, have POE-based guidelines covering a variety of functional elements. As discussed in the case study in chapter 9, the research laboratory generally lacks performance-based information and criteria. To date, very few POEs have been carried out on this complex and specialized building type.

HUMAN FACTORS

There has been extensive work in the area of human factors, which is also known as anthropometrics or ergonomics, the latter emphasizing work-environment-related human factors. This field is concerned with the dimensions and configuration of the designed environment, often the near environment, to match building occupants' physiological needs and physical dimensions. Equipment, such as a telephone, keyboard, or twenty-ton press in a manufacturing plant, will be designed with concern for human factors in terms of comfort, safety, and ease of operation.

An aspect of functional performance, the **human factor**, deals with the relationship between the **environment** and the physiological needs of the **building occupants**.

The greatly increased use of microcomputers and terminals at office workstations has led to extensive research of workstation flexibility to accommodate a variety of occupants. Seating is now designed to provide back support, as well as the need to change the body position over long periods of time. Environmental performance criteria for persons with a variety of handicaps have particularly improved since the mid-1970s and are now present in most building codes.

STORAGE

Not many criteria for storage exist, and those that do are often inadequate. Few POEs have been conducted in this area. Although the quantity of storage is a major criterion, the type, size, location, and distribution of storage are also important factors.

COMMUNICATION AND WORKFLOW

POEs and criteria exist for this element in the industrial sector and are beginning to be applied to the office environment. New technologies in both the factory and office are changing traditional design concepts. Examples are the office in the home in which employees use computers and, through modems, are "on line" with their companies.

FLEXIBILITY AND CHANGE

POE studies in hospitals documented the importance of varying amounts of flexibility in different environments. Studies in industry and offices also indicate that change is frequent in these building types and that rapid advances in technology make this functional element a high priority.

SPECIALIZATION WITHIN BUILDING TYPES

Building utilization and specialization are becoming important functional issues. Should buildings be equipped to accommodate any contingency, or should there be levels of sophistication within them? Segmentation and specialization within building types are becoming more frequent in office buildings, as well as in the hotel and retail industries.

Functional criteria responsive to different users, such as shoppers in retail development, include location, amount of square feet, level of finish, "mix" of stores, types of amenities, image, and even the configuration of the circulation path. Today, national franchises are using POE and market research to improve the performance of their various outlets.

Behavioral Elements of Building Performance

Behavioral elements of building performance are concerned with the impact of a building upon the **psychological** and **sociological** well-being of the building's **occupants.**

Behavioral elements of performance link occupants' activities and satisfaction with the physical environment. Behavioral evaluations are more profound—they are concerned with how the psychological and sociological well-being of building occupants is affected by building design.

How do the size of an area and the number of persons sharing it affect the building occupants? Does the functional distance between areas in a facility affect their frequency of use? Does the configuration of circulation routes affect social interaction? What features will best provide an appropriate image for a building? What design attributes provide for the occupants' perception of both an understandable and stimulating building? How can a satisfactory level of privacy as well as social interaction be developed for building occupants? These are some of the questions that are addressed by behavioral elements, and their physical design responses are rooted in the careful programming of buildings.

PROXEMICS AND TERRITORIALITY

Space is a basic concern in both architecture and the behavioral sciences. Spatial attributes, the sequence, location, relationships, shape, size, and detail of spaces have been shown to affect occupant behavior.

In recent years, research has progressed to include personal space and territoriality. Territoriality refers to a culturally set space, unlike the personal space occupied by the building user. Such territories are usually "controlled" by individuals or groups. This concept—the size of territory and its definition—has been found to be a particularly important factor in the design of housing, offices, and outdoor urban spaces.

Proxemics is the study of interpersonal distances maintained among individuals for purposes of communication. Such distances vary by culture, sex, activity, and age. A related concept, proximity, deals with the functional distances between building areas, their resulting frequency and quality of use, and social interaction.

PRIVACY AND INTERACTION

The development of this element of performance is logically connected with territoriality and proxemics. The control of access to an individual's or group's territory, including physical, visual, and aural access, defines the level of privacy or interaction that can be achieved.

Privacy or interaction can be experienced in various degrees, from an individual's private workspace to major urban spaces. Design elements that affect this attribute include the configuration of walls, openings, and access.

ENVIRONMENTAL PERCEPTION

How persons actually perceive buildings has a significant effect on design. Perception involves the senses in experiencing buildings. The Modern Movement of the early twentieth century, using gestalt theories of visual perception, created extremely simple and, to many persons, dull buildings. More recent theories indicate that persons can understand and show more interest in environments that are relatively complex and stimulating.

This performance element promises to have an important influence on building design. Buildings and spaces within buildings may have a defined range of performance attributes (scale, detail, size, color, lighting, and acoustics) appropriate to the needs of different occupant groups.

IMAGE AND MEANING

In addition to the physiological impact of a building, there is meaning attached to its design. Its shape, size, materials, details, and decoration form a kind of language, and just as a language conveys meaning, so does a building. This element of performance, long regarded as the realm of the critic and largely subjective, is undergoing a reinterpretation as the esthetic qualities of buildings are beginning to be measured through assessment of occupant responses. Although occupants' values and responses to buildings often differ from those of architects, a better understanding of these phenomena is critical for design decisions.

ENVIRONMENTAL COGNITION AND ORIENTATION

Information about buildings is interpreted and remembered by the user that may, or may not, be the goal of the client or designer. Environmental cognition is the sum of all of the above-mentioned behavioral elements of performance that help create mental maps in the minds of a building's occupants.

How does this multiplicity of information and elements become organized? One key principle is the creation of mental maps that aid in orientation. There are many large new buildings or buildings that have expanded over time in which finding one's way can be a primary design concern.

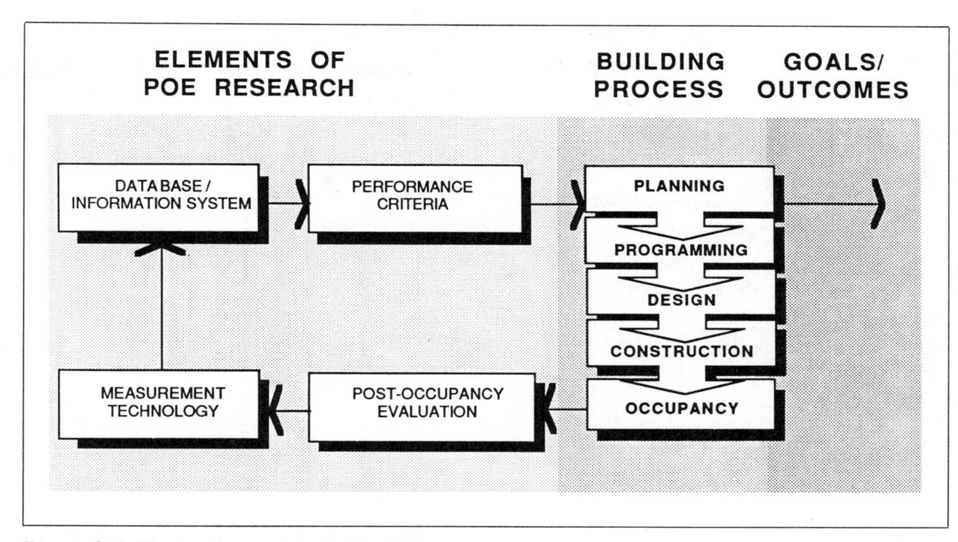

Figure 3-4. The performance evaluation framework.

THE PERFORMANCE EVALUATION RESEARCH FRAMEWORK

The performance evaluation research framework provides the linkage between the concept of building performance and the practice-oriented, actual POE phases and steps outlined in chapters 4 through 7.

The performance evaluation research framework (fig. 3-4) connects the evaluation of buildings with measurement technology, databases and information systems (including clearinghouses), and the development of performance criteria for buildings. This cyclic framework contains three important features: measurement technology, information systems, and performance criteria.

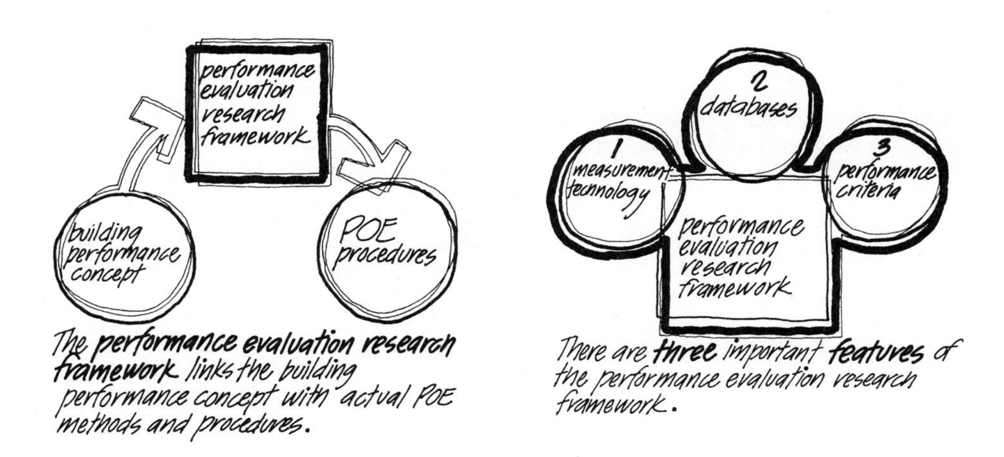

Measurement Technology

Measurement technology employs all those techniques and technological aids that are used in the data collection and analysis of POE. They include interviews, questionnaire surveys, direct observation, mechanical recording

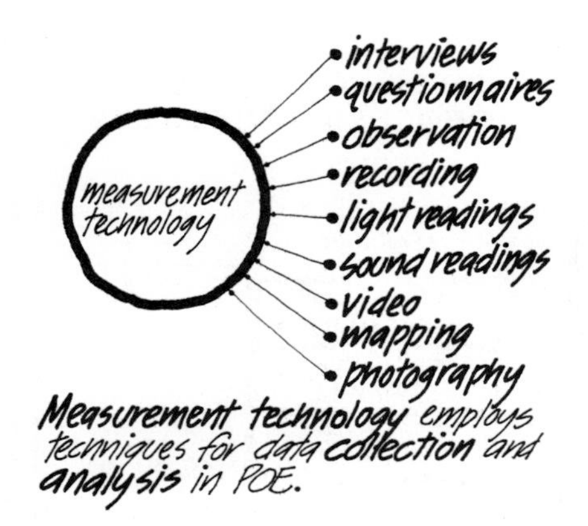

Measurement technology employs techniques for data collection and analysis in POE.

of human behavior, measurement of light and acoustic levels, recording with video and other cameras, and behavioral mapping.

To date, there is little standardization of measurement technology and methods in POE, posing long-term problems for this area. As more work is published, there will be the development of commonly used measurement methods.

Databases, Information Systems, and Clearinghouses

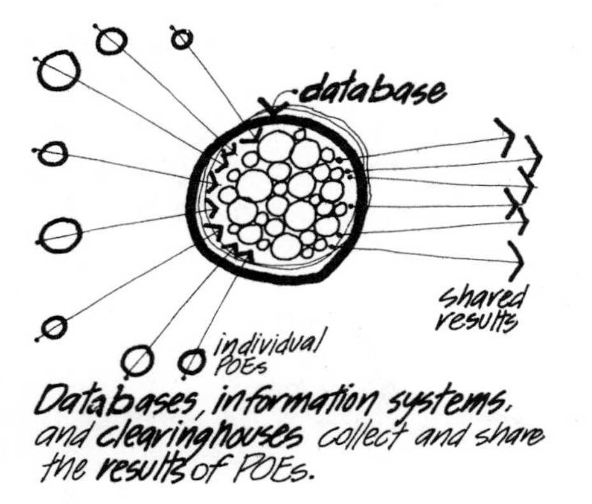

Databases, information systems, and clearinghouses collect and share the results of POEs.

Data collected with appropriate measurement technology is fed into databases, information systems, or clearinghouses that contain the results of POEs. These sources provide a much needed focus for sharing the POE results. The collection and dissemination activities are typically guided by organizations and associations concerned with specific building types, such as offices, schools, or housing.

Design guidelines are often distilled from a single author's or researcher's data and experience, while databases and clearinghouses tend to be broadly based and applicable to different user and building types. Such clearinghouses can also be accessed by practitioners and researchers for purposes of information exchange.

At this time, only one organized clearinghouse for the dissemination of POE research exists: the Architecture and Engineering Performance Information Center AEPIC (see appendix C). It was created to collect and disseminate information concerning technical failures in buildings. Its core of information comes from files donated by a major insurance company.

Other building-related information is also systematically collected and shared. For example, the Building Owners and Managers Association (BOMA) annually publishes *Experience Exchange Report* in both book and computer-disk form. Data on the annual income and operating expenses for hundreds of buildings are compiled and presented by location, building age, height, and size. Such POE-related clearinghouses will likely increase in number and include building programming information in the future.

Performance Criteria

Performance criteria are usually building-specific, address particular sets of occupants and functions, and are often documented in guides, manuals, or databases.

Performance criteria and guidelines are being developed from databases and other information systems for a given organization and/or building type. They are usually documented in either technical manuals, design guides, or specialized databases.

The criteria are building-specific and address particular sets of occupants and building functions. As such they are an evolving set of performance benchmarks for a given building type.

This chapter introduced the concept of building performance as a systematic process that compares explicitly stated performance criteria with the observed and measured performance of a building. Short-, medium-, and long-term implications of POE, when applied to the building process, were described, and elements of building performance were enumerated. Lastly, the Performance Evaluation Research Framework was presented, which links POE with measurement techniques and technology, databases and information, clearinghouses, design criteria and guidelines, and thus the building process as a whole. This chapter formed the theoretical foundation for the POE Process Model and its phases and steps, which follow in chapter 4.

Part 2

The Post-Occupancy Evaluation Process

POE is a comprehensive, hands-on process involving research but emphasizing the on-site examination of one or a number of buildings. An on-site POE, whatever its level of effort, includes real-world administration and management concerns, from estimating the cost of the project to obtaining access to all parts of the building to be examined. All of these techniques and concerns are addressed in chapters 4 through 7.

*The **POE process**, as described in this book, emphasizes both **survey research** and **evaluation of the physical environment**.*

4. The POE Process Model

While POE is a systematic and formal process, it can take place at different levels of effort—the same building can be satisfactorily evaluated in a day or two, a month or two, or even over several months.

The purpose of this chapter is to present a process model and an outline of steps for POE that can be applied to any type or size of building or facility, depending on the objectives of a given client organization and evaluation effort. The POE process model in figure 4-1 deals with the two major dimensions of POEs—the levels of effort and the major phases and steps involved in conducting a POE. Both dimensions are described in detail.

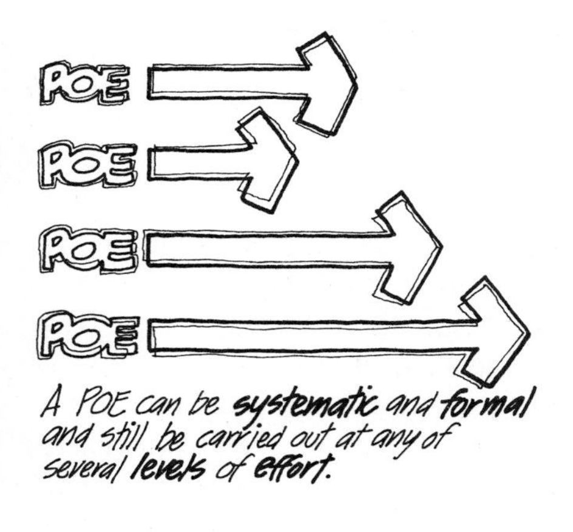

A POE can be **systematic** and **formal** and still be carried out at any of several **levels** of **effort**.

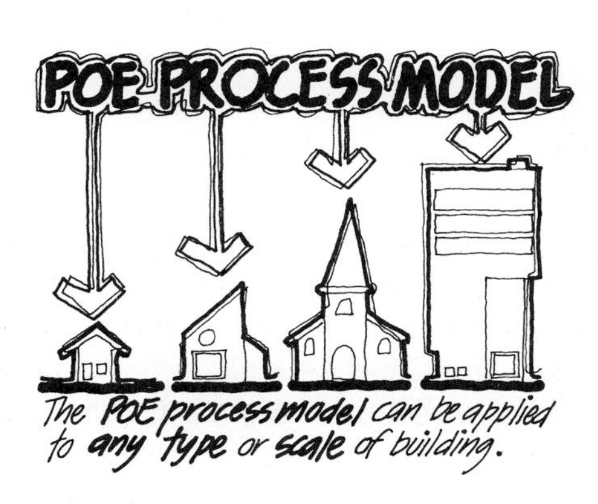

The **POE process model** can be applied to **any type** or **scale** of building.

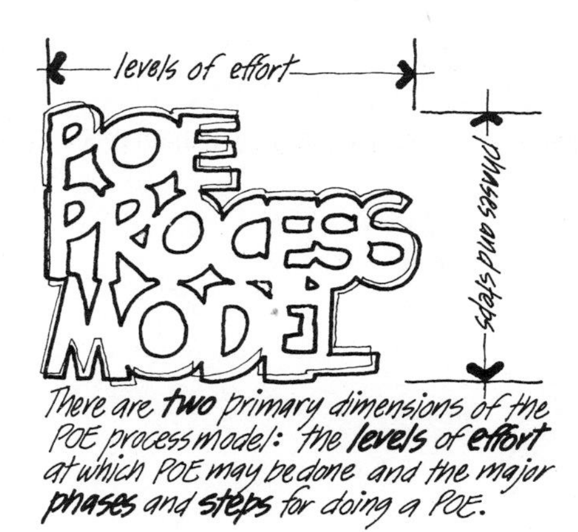

There are **two** primary dimensions of the POE process model: the **levels** of **effort** at which POE may be done and the major **phases** and **steps** for doing a POE.

THREE LEVELS OF EFFORT FOR POE

An analysis of the series of POEs that have been conducted by the authors since the mid-1970s and a review of the rapidly growing number of POEs

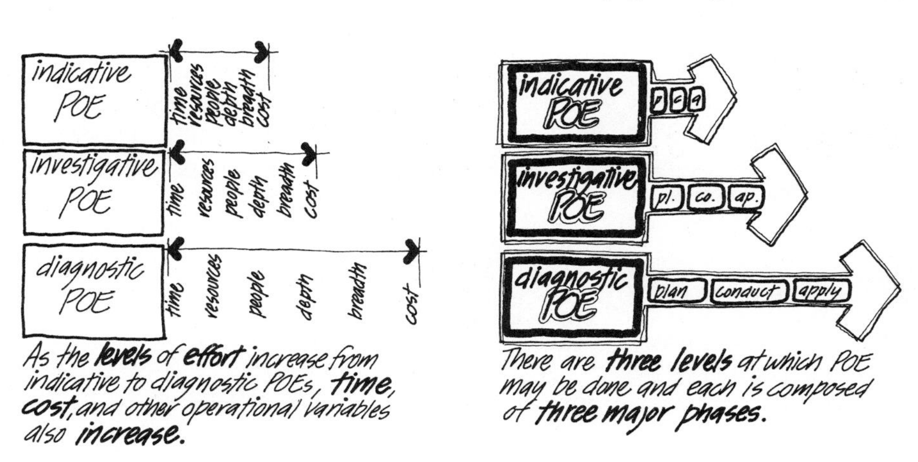

As the **levels** of **effort** increase from indicative to diagnostic POEs, **time**, **cost**, and other operational variables also **increase**.

There are **three levels** at which POE may be done and each is composed of **three major phases**.

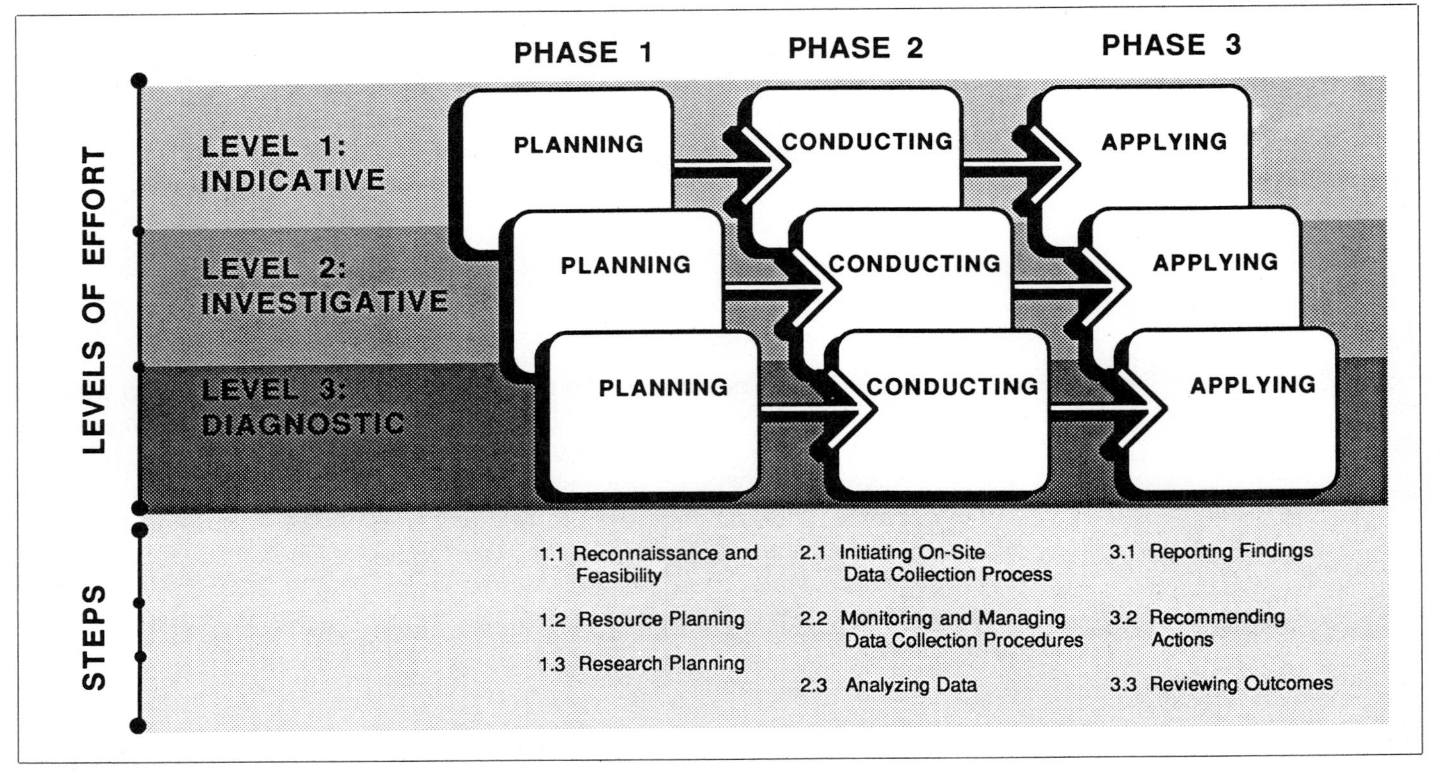

Figure 4-1. Post-occupany evaluation process model.

published have resulted in the emergence of three levels of effort for POEs, based on the amount of time, resources, personnel, the depth and breadth of evaluation, and, therefore, the implicit cost involved in carrying out POEs. These three levels—*indicative, investigative,* and *diagnostic*—each consist of three phases: planning, conducting, and applying the POE.

These three levels are distinct; they are not cumulative. The investigative-level POE, for instance, does not include indicative methods plus additional techniques. It is a qualitatively different research effort using methods that are appropriate at this level.

Level 1: Indicative POE

An indicative POE does what the name implies. It provides an indication of major failures and successes of a building's performance. This type of POE is usually carried out within a very short time span, from two or three hours to one or two days. An indicative POE presumes that the evaluator/evaluation

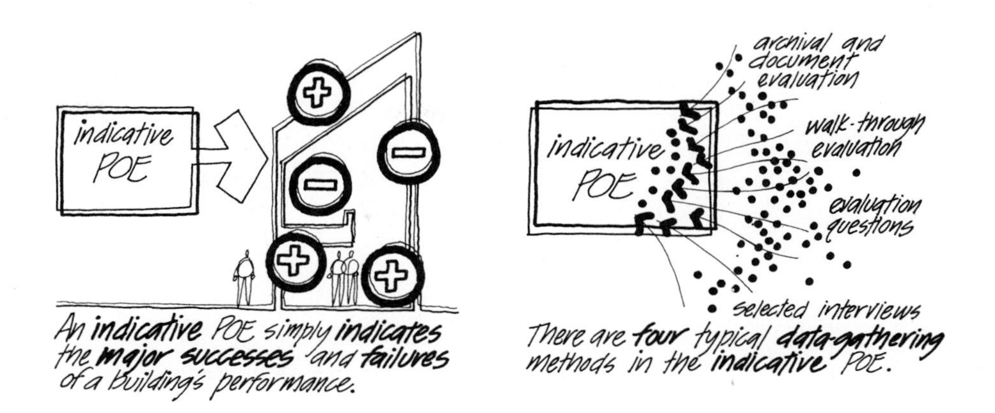

An **indicative** POE simply **indicates** the **major successes** and **failures** of a building's performance.

There are **four** typical **data-gathering** methods in the **indicative** POE.

team is experienced in conducting POEs and is familiar with the building type to be evaluated, as well as the issues that tend to be associated with it. The following data-gathering methods are typical of an indicative POE.

ARCHIVAL AND DOCUMENT EVALUATION

If possible, as-built drawings of the facility to be evaluated are obtained and analyzed. In addition, space utilization schedules, safety and security records, accident reports, remodeling and repair records, or any other historical/archival data that may be pertinent are obtained and analyzed. These POE activities do not necessarily occur on the building site.

PERFORMANCE ISSUES

A list of generic building-evaluation questions (see appendix B) is submitted by the evaluators to the client organization prior to the site visit. It is common that the facilities manager or committee delegated to deal with questions of space planning and building performance reply to open-ended questions concerning the performance elements. These questions deal with technical building performance as far as environmental conditions are concerned. In addition, they deal with functional appropriateness (adequacy of space and health, safety, and security issues, for example) and behavioral or psychological concerns such as the "image" of the facility. Replies to such questions represent management's knowledge not only of problems, but also of successful features of a given facility.

WALK-THROUGH EVALUATION

Following a discussion with management concerning the response to these performance issues, a walk-through evaluation is conducted, covering the entire facility and addressing the issues raised earlier. In addition, the evaluators use direct observation and, if warranted, still photography to identify building attributes that may deserve particular attention. Within a few hours, a walk-through can comprehensively cover a given building.

INTERVIEWS

Interviews with selected personnel responsible for and familiar with the facility and a debriefing of the client representatives conclude the on-site visit with the client organization. Subsequently, a brief summary of indicators of successful and unsuccessful features of the evaluated facility is submitted to the client organization for final verification and review.

Level 2: Investigative POE

An investigative POE is more time-consuming, more complicated, and requires many more resources than an indicative POE. Often an investigative POE is conducted when an indicative POE has identified issues that require further investigation, both in terms of a facility's physical performance and the occupants' response to it. The results of an indicative POE emphasize the identification of major problems. An investigative POE can cover more topics in greater detail and with more reliability.

While the major steps in conducting an investigative POE are identical to those in an indicative POE, the level of effort is higher, much more time is spent on the site, and more sophisticated data collection and analysis techniques are used. Unlike the indicative POE, in which performance criteria used in the evaluation are in part based on the evaluator's or evaluation team's experience, the investigative POE uses researched criteria that are objectively and explicitly stated.

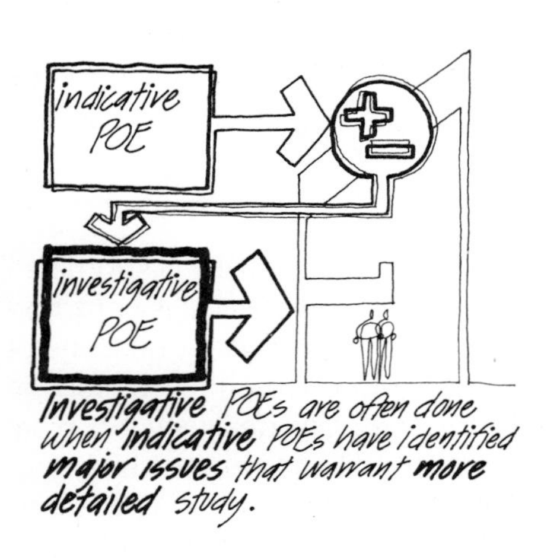

Investigative POEs are often done when indicative POEs have identified major issues that warrant more detailed study.

In an investigative POE, the evaluation criteria are explicitly stated before the building is evaluated.

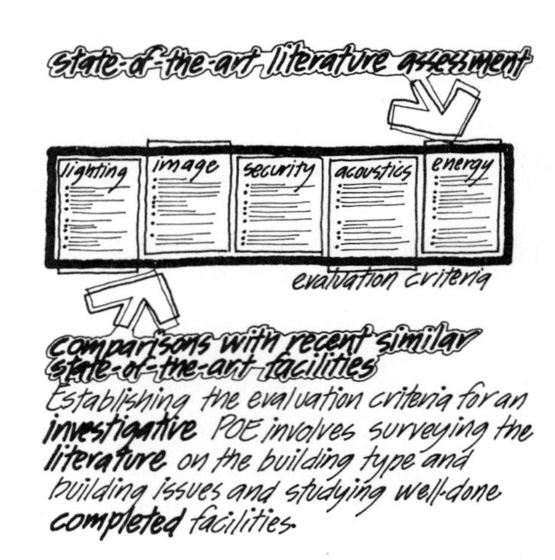

Establishing the evaluation criteria for an investigative POE involves surveying the literature on the building type and building issues and studying well-done completed facilities.

The establishment of evaluation criteria at the investigative level involves at least two types of activities: state-of-the-art literature assessment, and comparisons with recent, similar state-of-the-art facilities.

An investigative POE generally requires 160–240 man-hours, plus staff time for support services.

Level 3: Diagnostic POE

A diagnostic POE is a comprehensive and in-depth investigation conducted at a high level of effort. Typically, it follows a multimethod strategy, including questionnaires, surveys, observations, physical measurements—all approaches appropriate to the comparative evaluation of a cross section of facilities of the same type. The diagnostic POE may take from several months to one year or longer to complete. Its results and recommendations are long-term oriented, aiming to improve not only a particular facility, but also the state of the art in a given building type. Its methodology is very similar to traditional, focused research using the scientific paradigm.

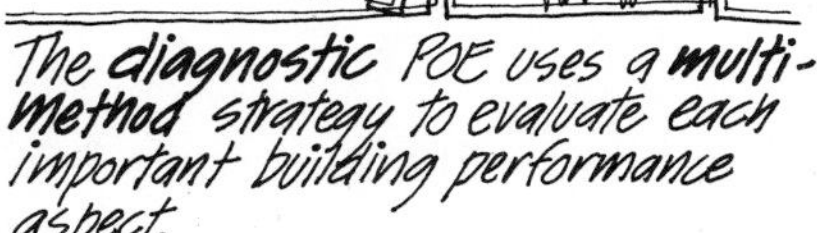
The **diagnostic** POE uses a **multi-method** strategy to evaluate each important building performance aspect.

The results of **diagnostic** POEs are meant to improve **particular** facilities **and** the state of the art in the **building type.**

The **methodology** used in **diagnostic** POEs is similar to that used in traditional **scientific research.**

Diagnostic POEs are usually large-scale projects, involving many variables. Often the attempt is made to develop results that indicate relationships among variables. Because of that, diagnostic POEs use sophistication in both data collection and analysis techniques exceeding that of investigative and indicative POEs.

An important part of diagnostic POEs has been research, the goal of which is the correlation of physical, environmental, and behavioral performance measures, thus providing a better understanding of the relative significance of various performance criteria. In chapter 1 a number of recent POEs of this type were noted (Marans and Spreckelmeyer 1981; Preiser 1983; Brill et al. 1984). Given all these prerequisites, diagnostic POEs hold the potential for making fairly accurate predictions of building performance and for adding to the state-of-the-art knowledge for a given building type through improvements in the design criteria and guideline literature.

OUTLINE OF POE PHASES AND STEPS

This outline of POE phases and steps is intended to be generic, to provide the reader with a basic understanding of each step's purpose, justification, activities, resources, and results, and, therefore, they do not necessarily apply to all POE projects, nor are all items listed under activities and resources needed or available in every project. (See also chapters 5 through 7.)

PHASES	STEPS	purpose	justification	activities	resources	results
plan	feasibility	•	•	•	•	•
	resource planning	•	•	•	•	•
	research planning	•	•	•	•	•
conduct	start data collection	•	•	•	•	•
	manage data collection	•	•	•	•	•
	analyze data	•	•	•	•	•
apply	report findings	•	•	•	•	•
	recommendations	•	•	•	•	•
	review results	•	•	•	•	•

*Typical generic POE **phases** and **steps**, together with the **purpose, justification, activities, resources,** and **results** of each are outlined in this section.*

Phase 1: Planning the POE

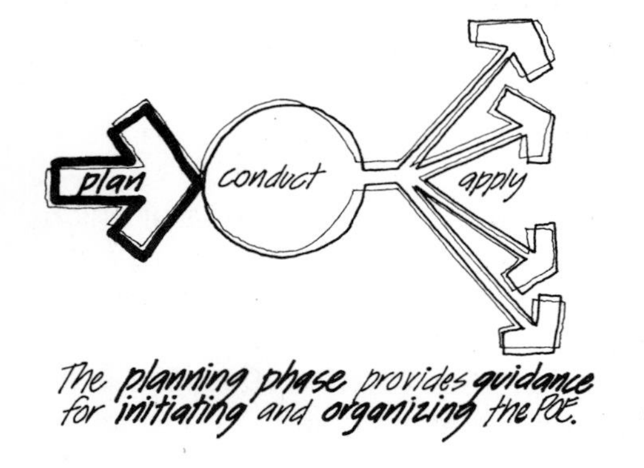

*The **planning phase** provides **guidance** for **initiating** and **organizing** the POE.*

There are several preliminaries to observe in initiating and organizing a POE project prior to the onset of on-site data collection. Liaison with the client organization is a critical aspect of all POEs. The client must be briefed on the nature of POEs, the types of activities involved, the resources needed, and client responsibilities in carrying out the POE. Agreement is reached on which type of POE to conduct. Then, historical and other background information that may assist in planning the evaluation is identified and obtained. Coordination with user groups within the building is begun and potential benefits for participants are outlined.

Resources for conducting the evaluation are organized, and a preliminary schedule, workplan, and budget are established in which project team members' tasks and responsibilities are defined. At the same time, appropriate research methods and analytical techniques are determined, and sources for evaluation criteria are identified. The three steps included in this first phase are reconnaissance and feasibility, resource planning, and research planning.

STEP 1.1: RECONNAISSANCE AND FEASIBILITY

a. *Purpose:* To initiate the POE project; to establish realistic parameters regarding the client organization's expectations of the evaluation; to determine the scope and cost of project activities; and to obtain a contractual agreement.

b. *Justification:* The development of a clear understanding of a POE—its process, information requirements, and client responsibilities—encourages a collaborative spirit between the evaluators and the client organization. Information about the building and the organizations housed in it, the past events, goals, and structure of the organizations involved, and the experience and concerns of key personnel, will assist in the initiation and organization of an effective POE. It will also help determine the scope of the project and resources necessary to conduct it.

c. *Activities:*
- Development of client contact
- Discussion of alternative levels of effort for POE
- Selection of appropriate POE level of effort
- Identification of liaison individuals
- Review of structure of client organization
- Reconnaissance of building to be evaluated
- Determination of availability of building documentation
- Identification of significant building changes and repairs
- Interviews with two to three key personnel
- Development and submittal of POE proposal for approval
- Execution of contractual agreement

d. *Resources:*
- Presentation of POE process through documents and slides
- Available precedents such as POE-based studies and criteria on the specific building type
- Evaluator's experience in POE and the building type
- The building in use
- Organizational structure
- Availability of key personnel
- Building documentation
- POE contract document

e. *Results:*
- Project proposal
- POE contract agreement
- Initiation of resource planning

STEP 1.2: RESOURCE PLANNING

a. *Purpose:* To organize the resources necessary for effectively conducting the evaluation, including reporting and applying results; to develop cooperation and support at all levels of the client organization.

b. *Justification:* The development of a management plan that includes the allocation of personnel, time, and money helps to ensure that project results are obtained on time. Concurrently, liaison personnel will assure the support and cooperation from many levels and groups in the client organization. This support will develop most easily if realistic objectives can be agreed upon.

c. *Activities:*
- Obtaining agreements from building occupants to participate in POE
- Definition of project parameters
- Development of workplan, schedule, and budget
- Presentation of resource plan to client organization

- Formation of POE project team
- Development of preliminary outline of final report

d. *Resources:*
- POE file on building type
- Building documents, plans, specifications
- Life-cycle documents: renovation, accident reports, and so on
- Client representatives / prospective interviewees
- Client organization's administrative procedures
- Contractor's past POE files or reports
- Project personnel
- POE methods and instrumentation
- State-of-the-art literature review

e. *Results:*
- Workplan for POE project team
- Budget breakdown
- Preliminary outline of final report
- Approvals for human subject involvement
- Initiation of research planning

STEP 1.3: RESEARCH PLANNING

a. *Purpose:* To develop a research plan that ensures that appropriate and credible POE results are obtained; to establish performance criteria for the building; to identify appropriate data collection and analysis methods; to develop appropriate instruments; to allocate responsibility for specific research assignments; and to devise quality control procedures.

b. *Justification:* Research planning provides the link between project resources and the quality or validity of the resulting POE process. Developing criteria for the performance elements is critical to facilitating a comparison between the actual building performance measures and the required conditions, that is, the performance criteria. Preliminary reconnaissance data are used to develop the overall research plan, including data collection, sampling, and analysis methods.

c. *Activities:*
- Identification of archival resources on client organization's documents
- Identification of prospective participants or respondents
- Contact with potential respondents in client organization
- Authorizations for photographs and surveys
- Presentation of outline of research plan to client
- Inspection of building
- Assignment of project tasks to available personnel
- Scheduling research tasks and personnel
- Development of research instruments
- Continued development of outline for evaluation report
- Classification and development of performance criteria for the evaluation

d. *Resources:*
- Computer-based information sources
- Design guidelines and standards of relevant government agencies and large building organizations
- Instruments and methods for data collection and analysis
- Consultants in, for example, research methods
- Current POE project file
- Availability of persons responsible for and familiar with the building
- Client organization's building-related files and documents

e. *Results:*
- History and description of building
- Diagrammatic drawings of building and rooms at various scales
- Data-recording equipment
- Technical, functional, and behavioral performance criteria
- Data-recording sheets
- List of respondents in client organization
- Preliminary schedule for on-site data collection
- Task assignments for project personnel
- Finalized research plan
- Analysis methods defined
- Performance criteria for building type
- Initiation of on-site evaluation

Phase 2: Conducting the POE

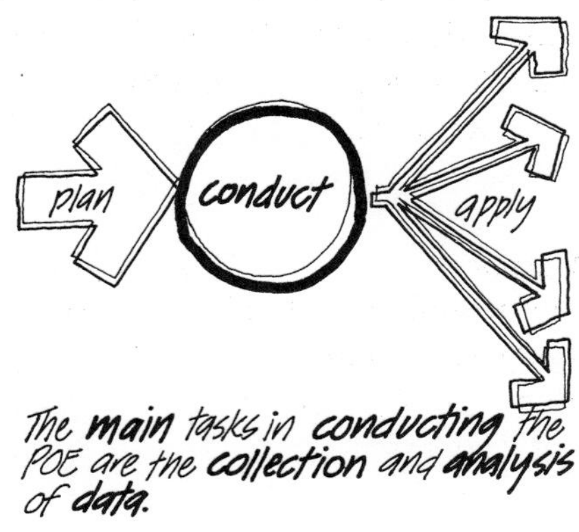

The main tasks in conducting the POE are the collection and analysis of data.

The activities, resources, and results involved in conducting the POE are based on those planning steps and tasks developed in phase 1, the planning phase. The main tasks in conducting the POE are the collection and analysis of data. Ensuring the quality of data collection and analysis, coordinating the many tasks involved, and continuing the liaison with the client are the primary objectives of this phase.

The steps involved in the conducting phase of POE are initiating the on-site data-collection process; monitoring and managing data-collection procedures; and analyzing data.

STEP 2.1: INITIATING THE ON-SITE DATA-COLLECTION PROCESS

a. *Purpose:* To prepare the evaluation team and the client organization for on-site POE activities; to coordinate the timing and location of POE activities in order to minimize disruption of the routine functions of the client organization.

b. *Justification:* Initiation of the POE includes both a logistical effort—moving to and locating on the POE site—and liaison with the occupants of the building. A smooth transition will aid in the coordination of evaluators and building occupants.

c. *Activities:*
- Coordination with building managers and users
- Building orientation for the POE team
- Practice runs of data-collection procedures
- Running a reliability check among observers concerning data collection
- Setting up of POE-team area in building
- Preparation of data-collection forms for distribution
- Preparation and calibration of data-collection equipment and instrumentation

d. *Resources:*
- Supplies and materials such as film, videotape, paper
- Equipment and instruments
- Copies of data-collection forms and other printed materials
- Access to copiers, phones, keys
- Identified respondents in the building

e. *Results:*
- Logistical support in place
- Final modifications to data-collection plan and procedures
- Final POE assignments
- Building occupants informed of on-site data collection
- Initiation of on-site data collection

STEP 2.2: MONITORING AND MANAGING DATA-COLLECTION PROCEDURES

a. *Purpose:* To assure collection of appropriate and reliable data.

b. *Justification:* The usefulness and reliability of the data—the actual building performance measures—largely depend on the care with which they are collected and recorded. Research planning, though critical to this effort, cannot anticipate difficulties, and, therefore, continuous monitoring of data collection is required.

c. *Activities:*
- Maintain liaison with client organization
- Dissemination of data-collection instruments such as survey forms
- Collection and collation of data-recording sheets
- Monitoring of collection procedures
- Documentation of POE process

d. *Resources:*
- Building maintenance staff
- Respondents in client organization
- Data-monitoring procedures and review process
- Instrumentation, data-collection forms and materials

- Research staff
- Consultants

e. *Results:*
- Raw data (measures of building performance)

STEP 2.3: ANALYZING DATA

a. *Purpose:* To analyze data; to monitor data-analysis activities in order to ensure reliable results; to develop findings that are useful and insightful.

b. *Justification:* Notwithstanding the earlier care in data collection, there may still be questions about the validity of certain data. After analysis is completed, a major task is the interpretation of the results to integrate the many findings into useful patterns and to indicate relationships among the factors examined.

c. *Activities:*
- Review of reliability of raw data
- Data entry and aggregation
- Data processing
- Review of results of data analysis
- Interpretation of data
- Development of findings
- Structuring of results
- Completion of data analysis

d. *Resources:*
- Data-analysis procedures, software, and equipment-performance criteria
- Data analysis consultants
- Research staff
- Data-interpretation aids—graphics, classification, schemes, and so on

e. *Results:*
- Data analysis
- Findings
- Interpretation of data
- Initiation of application phase

Phase 3: Applying the POE

In this phase of the POE, findings are reported, conclusions drawn, recommendations made, and eventually the resulting actions reviewed. The findings of the POE are organized, and an effective reporting framework is devised. Usually, recommendations imply that the results must be prioritized, a task requiring continued liaison with the client. Finally, actions resulting

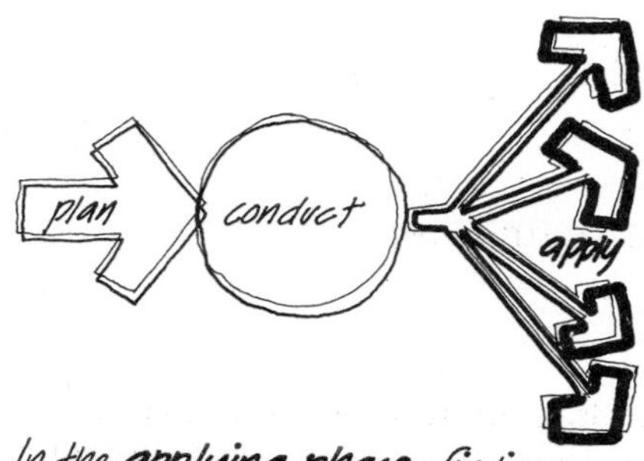

*In the **applying phase**, findings, conclusions, and recommendations are formulated together with a method to **review the results** of actions taken.*

from the POE are reviewed to ascertain that benefits envisaged in initiating the evaluation have, in fact, been achieved.

The steps involved in the applying phase of POE are reporting findings; recommending actions; and reviewing outcomes.

STEP 3.1: REPORTING FINDINGS

a. *Purpose:* To report the findings and conclusions of the POE in a way that is appropriate to the client's needs and expectations; to provide clear and accurate data that support the findings and recommendations.

b. *Justification:* Reporting and appropriately presenting the findings of the POE are critical for the client's understanding of the results. This increases the likelihood that the client, as well as other relevant organizations and professionals, will use the information presented.

c. *Activities:*
- Preliminary discussion of findings with client
- Development of presentation formats
- Organization of report contents and other presentations
- Preparation of documentation
- Formal review of findings by client organization
- Dissemination of reports

d. *Resources:*
- Client and liaison personnel
- Current POE project file
- Previous POE files or reports
- Research staff
- Graphic-design equipment and supplies
- Editorial and graphics consultants

e. *Results:*
- Documented POE information
- Approval of report by client
- Published final report
- Disseminated final report
- Implementation of recommendations

STEP 3.2: RECOMMENDING ACTIONS

a. *Purpose:* To make recommendations in order to implement feedback and feedforward; to stimulate action based on the findings and conclusions of the POE process.

b. *Justification:* The development and prioritization of recommendations require continued discussion and analysis with the client. Alternative strategies are developed, and the costs and benefits of each are examined. This step ensures that the most appropriate actions for the client are initiated.

c. *Activities:*
- Review of project findings and needs with client and building occupants
- Analysis of alternative strategies
- Prioritization of recommendations
- Actions for implementation of recommendations

d. *Resources:*
- Client organization's facilities, operations, and management
- Analysis techniques for prioritizing recommendations
- Research staff
- Final project report

e. *Results:*
- Approval of prioritized strategies and recommendations
- Recommended actions
- Identification of the need for additional research in certain topic areas

STEP 3.3: REVIEWING OUTCOMES

a. *Purpose:* To monitor the life-cycle implications of the recommendations.

b. *Justification:* The POE should result in improved performance of the subject of the POE and subsequent buildings that have been changed based on the POE. Monitoring the performance of such buildings affirms the integrity of the POE process and verifies direct benefits to the client.

c. *Activities:*
- Liaison with client organization
- Continued review and monitoring of implemented recommendations
- Reports on results of the effects of changes to the evaluated buildings and subsequent buildings

d. *Resources:*
- Client liaison
- Current POE project file
- Final POE report
- Instrumentation and surveys

e. *Results:*
- Completed project file
- Dissemination of the effects of POE-based design to the client organization, design professionals, and owners and facilities managers in general

This outline of the POE process indicates the research, administrative, and logistical concerns that even an indicative POE should consider. An underlying assumption of the POE process is that it is formal and rigorous so that, gradually, real progress in building performance can result.

5. Planning the POE

Planning the POE involves three Rs: reconnaissance, resource planning, and research planning:

1. *Reconnaissance and feasibility:* Determining the scope and level of effort will help in selecting the most appropriate type of POE—indicative investigative, or diagnostic.
2. *Resource planning:* What methods of data collection and analysis will work best for a given POE level of effort? How should personnel be allocated? How long should the individual tasks take? In this planning task, one chooses appropriate methods and personnel, and the POE is scheduled in detail.
3. *Research planning:* This activity is the equivalent of working drawings in design. The research plan is the final step before conducting the actual on-site POE. All data collection and analysis methods are decided upon, as are the techniques for ensuring the quality of the data.

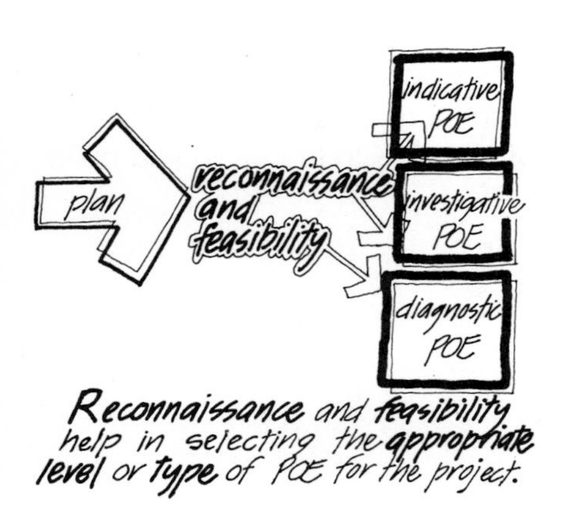

Reconnaissance and feasibility help in selecting the appropriate level or type of POE for the project.

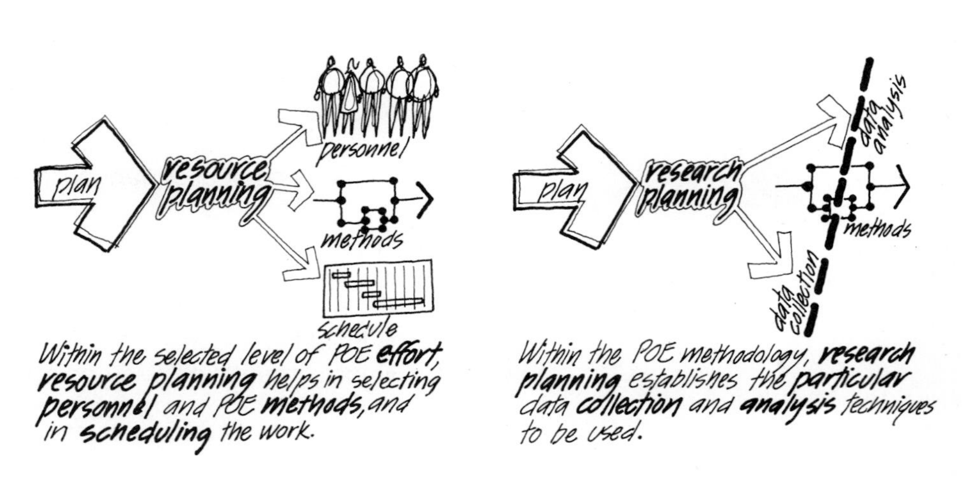

Within the selected level of POE effort, resource planning helps in selecting personnel and POE methods, and in scheduling the work.

Within the POE methodology, research planning establishes the particular data collection and analysis techniques to be used.

Planning a POE includes both administrative and research responsibilities. At this time the scale of most POE studies is relatively small, and the managerial and research functions can be combined, enhancing project coordination.

RECONNAISSANCE AND FEASIBILITY

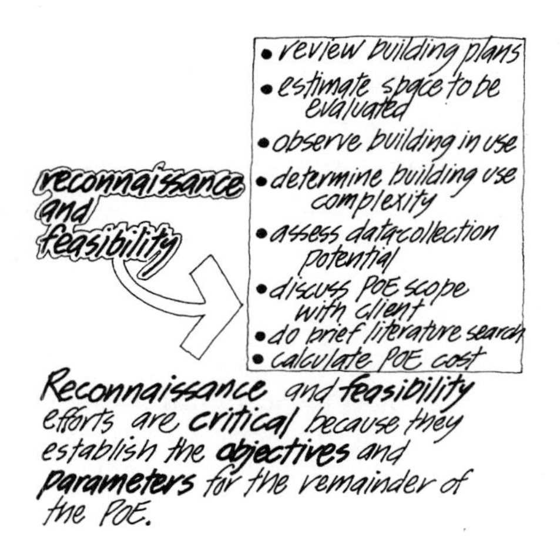

This critical first step establishes the objectives and parameters for the remainder of the POE. A number of factors must be assessed before determining the scope of the project. Architectural plans should be obtained for the building to be evaluated. A preliminary estimate of the number of rooms and the level of complexity of the building, including its mechanical systems, if appropriate, should be made.

The building should be observed under working conditions. Floor plans never fully represent the complex network of activities going on in a building. A few hours of observation should suffice in making a preliminary judgment concerning the intensity and complexity of building usage. The ability to collect data, particularly regarding behavioral performance measures, should be assessed.

Discussions with the client should clarify the goals, scope, and depth of the POE. A brief search of the literature should be undertaken, focusing on a review of the state of the art for the building type in question. Relatively few building types have been the subject of substantial POE research, and therefore, experts on the building type being evaluated should be consulted.

The budget is the bottom line. If the client's goals and desired scope of the POE exceed the budget, appropriate compromises must be made.

Selecting the Appropriate Level of Effort

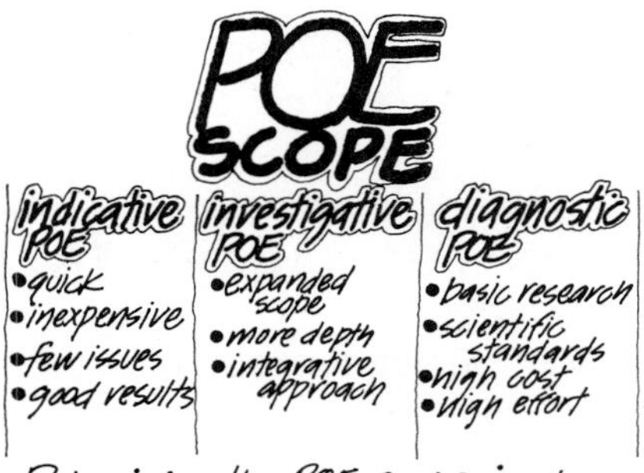

Determining the POE scope involves choosing one of the three POE levels of effort.

Chapter 4 categorized and described the three levels of effort of POEs. The objective of the reconnaissance phase is to determine which one to use.

The indicative POE may be quick and inexpensive, but it can yield good results, particularly when focused on a few evaluation issues. When budgetary constraints are critical, this indicative or walk-through type of POE is appropriate. On-site activities can take from a few hours to a few days, depending on the size of the building. More time may be spent preparing for or following up on the walk-through POE than carrying out the on-site activities themselves. This preparation or follow-up time involves becoming familiar with the building and building type and analyzing and writing up the results in a short report or technical letter after the on-site visit.

Many POEs are carried out at the investigative level of effort. This type of POE covers more ground, more thoroughly, and with better reliability than the indicative POE. The investigative POE is typically an integrative approach; it is comprehensive, although certain elements of the building performance may be excluded. While many specific performance elements are examined, the results are organized into fewer and larger categories than in the diagnostic POE. The depth of investigation is careful (although not exhaustive); the results tend to be of good quality and can be used to make major decisions about improving the performance of the building in question.

The diagnostic POE represents a major, basic research effort that uses significant resources and is conducted according to state-of-the-art standards. Weighted toward basic research, this type of POE is usually sponsored by large corporations or government agencies, often at the federal level. Although they are relatively high-cost, high-effort studies, diagnostic POEs are cost-effective for organizations that construct certain building types on a repetitive, large-scale basis.

The Cost of POE

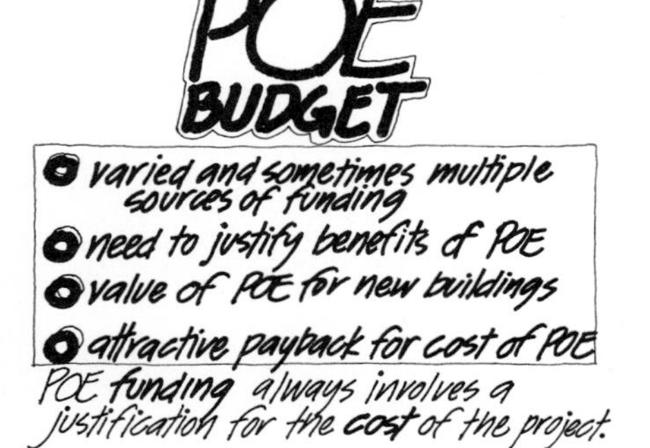

Funding, as in most other endeavors, remains the key determinant of the effort and resources invested in a POE. Funding and support for a POE can come from a wide range of sources. In the design of new facilities, POEs are seen as investments. The relatively small cost of a POE will be paid back many times during the life of the building and in improvements to future buildings of the same type.

The payback or benefits of a POE are considerably leveraged, particularly for a long-term building user. Whether energy is saved, maintenance and repair costs are reduced, shoppers are encouraged to spend money, future changes are facilitated, crime is lessened, or productivity and satisfaction are raised, net savings are the result. This leverage factor, where the benefits of a POE greatly outweigh its cost, has even more significance when a POE is used as input to a series of buildings.

Thus far, there has been no thorough documentation or analysis of the costs of POEs. The estimates provided here are based on the authors' experience in POE studies conducted at the three levels of effort. These estimates are based on 1986 costs and should be modified by the scope, size, complexity, and reliability required of the POE.

- An indicative POE should cost approximately 5 to 10 cents per square foot of evaluated floor area (for a 40,000–square-foot school, an indicative POE would cost $2,000 to $4,000).
- An investigative POE has a cost range of 15 to 50 cents per square foot (an investigative POE for the same school would cost between $6,000 and $20,000).
- A diagnostic POE costs approximately $1 per square foot (the cost for the same school would be upwards of $40,000).

RESOURCE PLANNING

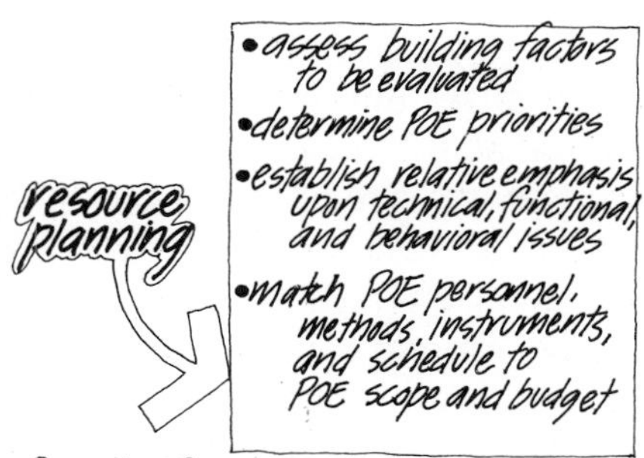

*Once the POE level of effort has been chosen, **resource planning** determines what is **required** to **implement** the POE.*

Once a level of effort has been determined, resource planning can proceed. It involves allocating project personnel, scheduling the entire process, and developing a project budget.

Within any of the three basic effort levels of evaluation, a range of resources and techniques is at the disposal of the evaluator. Are technical, functional, and behavioral factors all to be assessed? Are any of these topics to be given special weight? Are there specific elements that should be emphasized? Again, answering these questions sets the stage for the entire POE. During the resource-planning phase of the POE, the personnel, timing, methods, and instruments must be matched to the predetermined POE scope and budget.

Personnel

*Usually at least **half** the cost of a POE will be allocated to the **wages** and **benefits** of the POE **personnel**.*

The level of effort of a given POE is directly related to the cost of personnel working on the project. At least half of the cost of a POE done by a professional consultant, whether business- or university-based, will be allocated to the POE team. It is critical to have experienced evaluators; less experienced personnel should be given only basic responsibilities in a POE, such as certain data-collection tasks. The client should also understand that POEs are still a rather new process and that, although some procedures are standardized, there still needs to be much customization in order to respond to specific evaluation circumstances.

Time

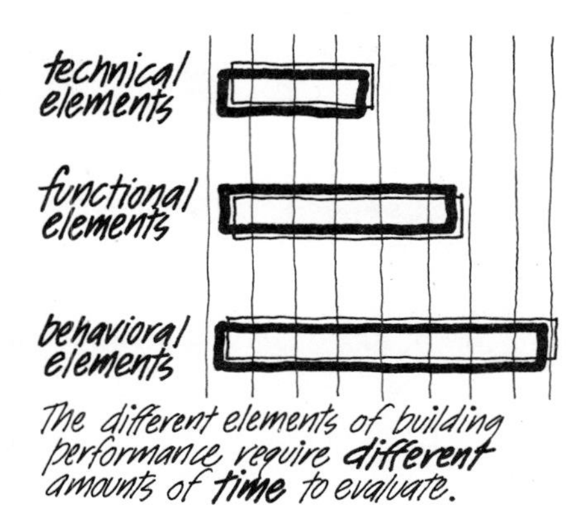

*The different elements of building performance require **different** amounts of **time** to evaluate.*

Different elements of a POE take very different amounts of time to evaluate. Specific technical elements can be evaluated in a relatively short time, excluding those phenomena that occur only under seasonal or other special conditions—rain driven from various wind directions, for example. A walk-through, indicative type of POE is most appropriate in this case.

Functional performance elements take somewhat longer to evaluate. Most organizations and occupancy types have a considerable amount of routine usage that can be investigated relatively easily. Nonroutine use, however, can be the most important function to a client organization. For example, Christmas traffic at an airport or shopping center, employee or function relocations in an office building, or temporary production crises are all critical in determining the appropriate levels of service and capacity.

The evaluation of behavioral performance elements is most labor- and time-intensive. Building users, settings, and activities exhibit considerable variety, even in seemingly small buildings or traditional building types. Activities also change by season and over time, for example, during a college semester. Because of such changes, data on the behavioral performance elements of a POE often need to be collected over a period of time that is representative of typical building usage, including cyclical activities as well as less frequent special events.

Methods and Instruments

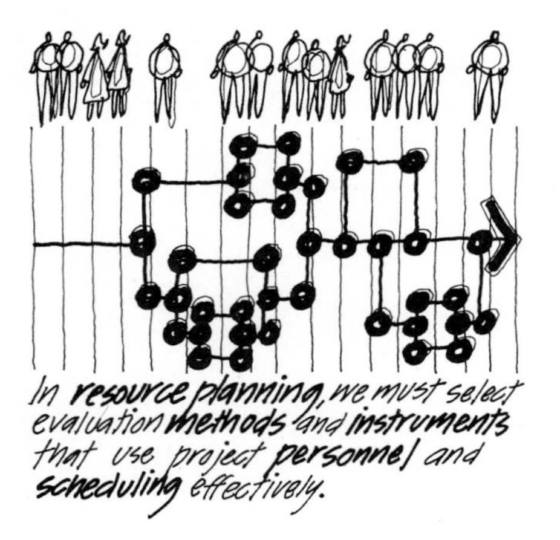

In resource planning, methods must be selected so that project scheduling and personnel time are effectively used. The research plan deals with the specifics of methodology, sample size, and analytical techniques, thus requiring coordination of the resource and research plans.

Consideration of appropriate methodology is most critical when evaluating behavioral performance elements because of the extensive amount of time that is required for both data collection and analysis. When feasible, a multimethod approach is strongly suggested to enhance the credibility of findings. For instance, photography may be combined with observation, and a questionnaire may be followed by focused interviews. If available, archival records may be correlated with data collected by other methods. Because a multimethod data-collection strategy uses more resources, it is better to focus on a few critical behavioral performance elements. For example, in shopping facilities, this may mean an emphasis on image and circulation patterns, or in housing, on symbolic elements, and in offices, on workflow and communication.

RESEARCH PLANNING

With schedules drawn, budgets fixed, and responsibilities assigned, a research plan can be developed.

A research plan has five components.

- Determining the critical items to be investigated
- Choosing indicators that will represent these items
- Developing specific measures for the indicators
- Setting criteria to evaluate the measures
- Anticipating results and conclusions

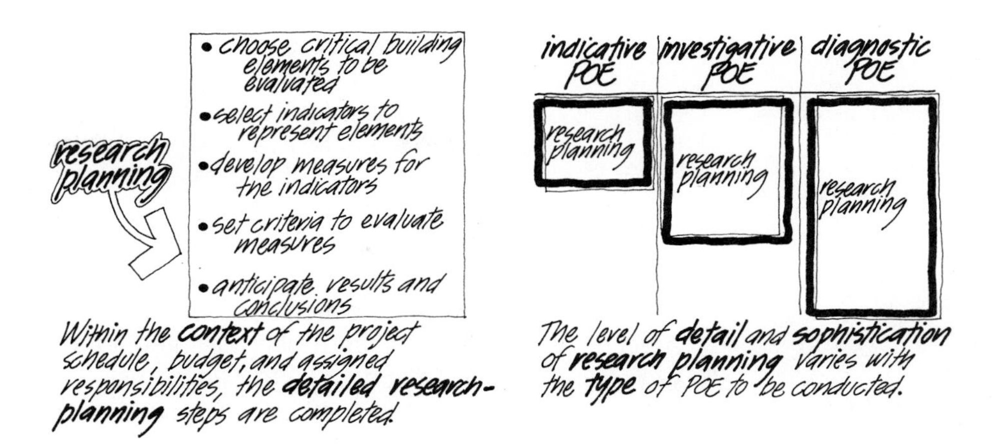

Within the **context** of the project schedule, budget, and assigned responsibilities, the **detailed research-planning** steps are completed.

The level of **detail** and **sophistication** of **research planning** varies with the **type** of POE to be conducted.

The differences in the three levels of effort in POEs are very apparent at this point. An indicative POE is a systematic and well-defined walk through the building, making observations without further testing or development of explicitly stated performance criteria. Research planning and preparations for indicative POEs are straightforward and simple.

An indicative POE typically places equal emphasis on the planning, conducting, and applying phases. Building plans and specifications are obtained and analyzed before the walk-through tour. Reduced plans of the building are used for note-taking during the course of the investigation, with annotations entered into respective spaces and building areas on the plans.

A standard form or checklist helps to inventory all building areas and elements of performance to be investigated. In its most simple form, this list can contain major evaluation categories amounting to as many as one hundred items: from lighting, interior wall surfaces, and storage to privacy and other concerns. At this level of effort, measures can take the form of notes on the plans and checks on the rating scales, lists, and so on.

Indicative POEs are based on existing benchmark examples of the building type being evaluated. Critical in such evaluations is the evaluator's familiarity with exemplary buildings rather than explicitly stated criteria.

The research plan for an indicative POE is often based on the ultimate use of the POE. For example, in an office building with repetitive floor plans and similar functions, not all floors may need evaluation. However, if the indicative

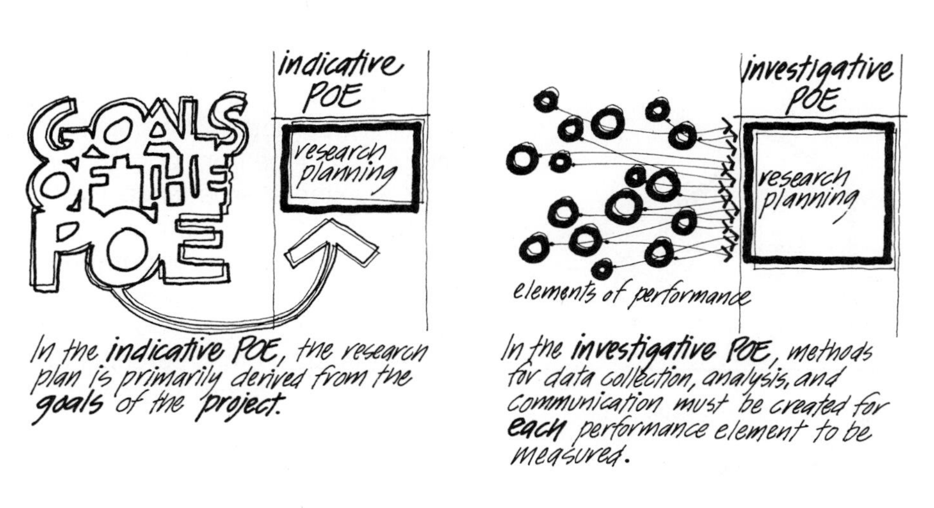

In the **indicative POE**, the research plan is primarily derived from the **goals** of the **project**.

In the **investigative POE**, methods for data collection, analysis, and communication must be created for **each** performance element to be measured.

POE will be used for legal purposes such as an appraisal or a building condition report, the entire building must be carefully examined.

An investigative POE requires considerable time for preparation of a research plan. Systematic evaluation at this level means that appropriate data collection, analysis, and communication methods and formats will be created for each performance element to be measured. (See appendix B and chapter 9.)

Measurement instruments should be obtained and arrangements for calibration made. Cameras and hand-held tape recorders should also be used for data collection.

The research plan for the diagnostic POE will resemble the investigative POE in concept, but it has considerably more depth and breadth. Triangulation or multimethod research, to ensure the credibility of findings, is standard in a diagnostic POE. Data on more discrete variables are collected, and analysis methods are much more sophisticated than at the indicative level.

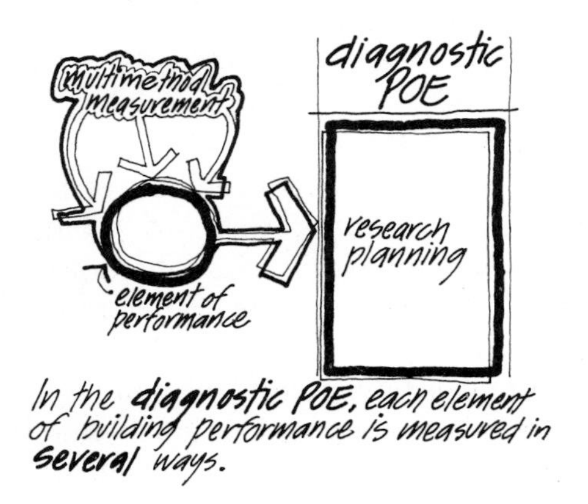

*In the **diagnostic POE**, each element of building performance is measured in **several** ways.*

An early attempt to describe evaluation techniques for designers was made by Sanoff (1968). Space does not permit a detailed description of research design and methods of data collection and analysis. Presented below are certain aspects of measuring user responses that are deemed important. For more information on these topics, see: Zeisel (1981) for overall evaluation strategies; Survey Research Center of the University of Michigan (1976) for interview techniques; Oppenheim (1966) for aspects of questionnaire design; Kish (1965) for survey sampling; Webb et al. (1966) for nonverbal response measurement methods; and Nie et al. (1970) for an example of statistical packages for purposes of data analysis.

Rating Scales

Considerable discussion in the research literature has focused on the value of data obtained through rating scales and the precision that is associated with them. Rating scales presume that a perceived continuum exists in the respondents' minds that permits an assessment of building performance ranging from very negative to neutral to very positive. For example, there is disagreement among evaluators over how many points there should be on rating scales. Some evaluators favor a seven-point or even an eleven-point scale, while others maintain that a four- or five-point scale is adequate. The advantage of the four-point scale is that because it has no neutral midpoint, it forces respondents to commit to a positive or negative assessment of the building-performance element being evaluated. Simple, four-point rating scales such as those depicted in the questionnaire (fig. 9-7) provide for performance ratings that range from poor to excellent.

In the authors' experience, the longer, therefore more differentiated scales do not provide any particular advantage over short scales, since data are aggregated and averages computed for the responses, thereby obviating the differentiations the longer scales afford. It has also been found that excessive "number crunching" is not worth the effort; the aim of the rating scales is simply to obtain gross perceptions of building performance. These perceptions are highly subjective and probably variable according to the circumstances affecting the respondents' judgments at a given moment. In fact, data

*Building user **rating scales** usually measure only **gross** perceptions of **building performance**.*

generated from rating scales are indicative in nature and point to issues that may be pursued further through other methods of data collection, such as in-depth interviews.

Semantic Differential Scales

The semantic differential, while developed in the field of linguistics for measuring the meaning of words, has been extended and sometimes misused by environmental psychologists in an attempt to assess visual and other performance elements of buildings. Two shortcomings can be mentioned in this regard: First, the meanings of descriptive adjectives that are used in bipolar rating scales are not always shared among the raters. And, second, once factor analyses have distilled clusters of descriptors (so-called dimensions), it is not always easy to determine the specific physical attributes of the building under investigation to which these "dimensions" refer.

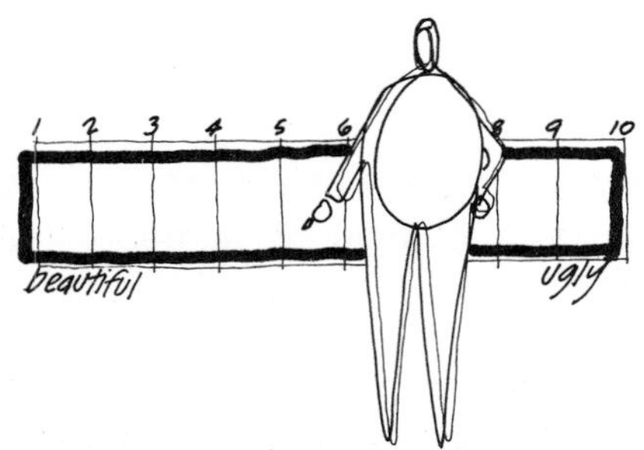

*Semantic differential scales are often used to measure **visual** and other performance elements of buildings.*

Aggregating Qualitative versus Quantitative Data

In cases where the qualitative performance aspects of a building are the focus of the POE it is often advisable to avoid aggregating data, to seek to isolate those responses that disagree with the majority, and to follow up on the reasons for this disagreement. In assessing visual esthetic quality, for example, social, cultural, regional, and other biases are significant differentiators. The disadvantages of aggregate data are that, through the computation of means or averages, important, fine differences in the data are eliminated, thus giving an incorrect assessment of the phenomena that are present.

Often, several indicators are used to evaluate major phenomena. Observers are carefully trained and evaluated, and truly representative, stratified samples are used in data collection. Sophisticated methods of data analysis are emphasized. Cross-tabulations of variables are typical, and inferential statistics are planned for and applied.

Research Planning by Category of Research

TECHNICAL FACTORS

For the experienced evaluator, these are the least complicated elements to measure, even at the diagnostic level. Indicative measures are based primarily on observations, while investigative and diagnostic POEs use instrumentation whenever possible. Diagnostic evaluation is done quite rigorously, and automatic recording devices are often used to obtain complete data over a

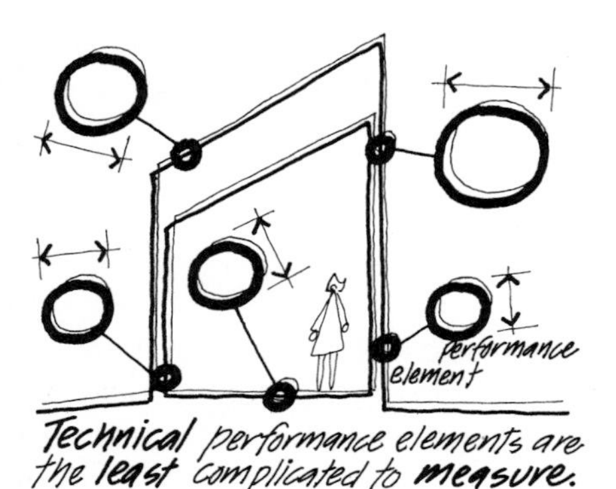

*Technical performance elements are the **least** complicated to **measure**.*

period of time, for example, on temperature changes and acoustical performance. Diagnostic POEs, however, do not quite approach the level of detail common in traditional, focused building research, for example, on energy-conservative design or structural failures.

Analysis at the diagnostic level of effort may combine several categories of research. Technical and behavioral elements of performance may be examined together, such as the use of energy-saving devices and the occupants' control of these devices, the effect of acoustical qualities on user performance in open plan schools or offices, or the effect of task lighting on productivity.

FUNCTIONAL FACTORS

In an indicative POE, the evaluation of functional factors usually occurs at a very basic level. The presence, frequency, and location of elements that support activities are determined and compared to the expert evaluator's experience with similar benchmark facilities.

Investigative and diagnostic POEs develop performance criteria for the activities and functions in the building. An investigative POE will monitor specific functions over time, compare them to existing criteria, and evaluate their performance. A diagnostic POE, on the other hand, in evaluating these same factors, will question existing criteria and independently develop performance criteria for some elements based on actual use.

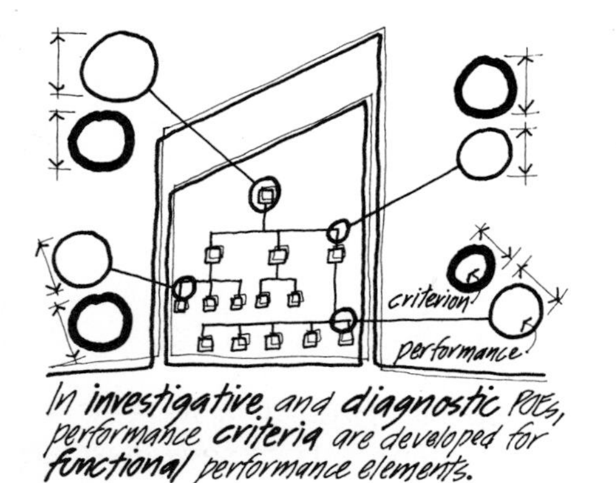

*In **investigative** and **diagnostic** POEs, performance **criteria** are developed for **functional** performance elements.*

BEHAVIORAL FACTORS

There are three important differences between investigative and diagnostic POEs in measuring behavioral performance elements. Diagnostic POEs advocate:

- The use of multimethod approaches in data collection and analysis, providing a high level of credibility.
- More extensive data collection and sophisticated statistical analysis techniques are employed.
- An emphasis on basic research, attempting to probe correlations and causalities.

*In measuring **behavioral** performance elements, the **diagnostic** POE research plan is different from the **investigative** plan in several ways.*

The case study in chapter 10 provides examples of research planning for a diagnostic POE.

Assiduous planning creates the foundation for the quality of POEs. The reconnaissance phase establishes the scope and parameters for the entire project. The resource-planning steps determine the allocation of time, funding, and personnel, and the research plan itself lays out the data-collection as well as data-analysis procedures.

6. Conducting the POE

While the success of the POE depends largely on the careful and structured preparation of the planning phase, carrying out these plans is the major event and responsibility in a POE.

A key task in conducting the POE is data collection. As data are collected, there is the large perspective to be considered. A great deal can be learned by experiencing firsthand the environment under evaluation. Combined with the analysis of data, this provides an analytical and creative basis for developing deeper insights into the environment—insights that might not be obtained from the analysis of data alone.

Data collection is a key task in *conducting* the POE.

Data collection and *analysis* should always be combined with *direct experience* of the evaluated environment.

Conducting the POE can be thought of in three steps:

1. Initiating the on-site data-collection process, which includes mobilization of all parties involved in the POE
2. Monitoring and managing data-collection procedures, which includes establishing practical guidelines
3. Analyzing data, which includes data interpretation

1. INITIATING THE ON-SITE DATA-COLLECTION PROCESS

Officially, the POE begins when an agreement is reached between the client and the researchers/evaluators. In many people's minds, however, the POE actually begins when the evaluators enter the building and begin to examine it. The on-site portion of the study is often the most visible stage of the POE.

Initiating the on-site POE is essentially a mobilization effort. While it does not include the actual collection of evaluation data, this step does include establishing a base of operations for the POE team and pilot-testing and coordinating data-collection methods.

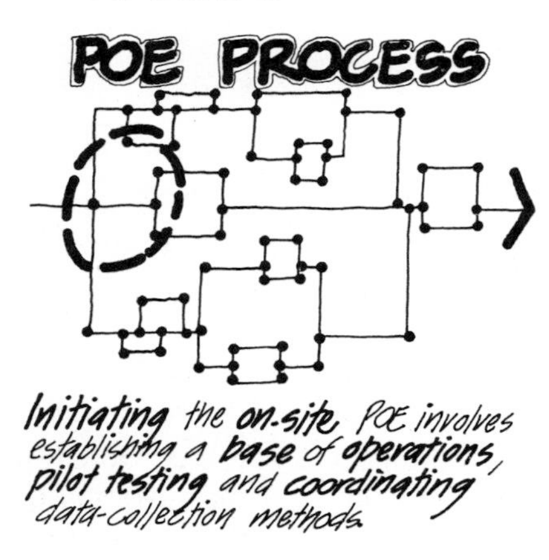

Initiating the on-site POE involves establishing a base of operations, pilot testing and coordinating data-collection methods.

Mobilization

Mobilization is an administrative aspect of POE. While it may have little intellectual content, it must be performed efficiently so that the evaluation team can focus on the main task without distractions or wasted time. There are many such administrative tasks in every POE.

Mobilization is especially important in investigative and diagnostic POEs. In indicative POEs, little or no equipment is used, few forms are required, and no special work area for the evaluation team may be required. The time spent in the building is usually short, and most of it is on the go.

At the investigative and diagnostic levels of POE, mobilization and field headquarters are required. Usually, the same instruments are needed for both these types of POEs. Instruments refer to the activity of data gathering, for example, through survey questionnaires and structured interviews that address a standard set of questions, or a standardized set of observations that permit the discovery of patterns of behavior of the occupants of a given facility.

For investigative and diagnostic POEs, it is important to have a POE field headquarters somewhere in or near the building to be evaluated.

The Reactive Effect

Great care needs to taken by POE researchers to minimize the effect of an impending study on the occupants of a facility. The so-called Hawthorne effect (Roethlisberger and Dickson 1939) is a good example of a situation where the mere occurrence of a study or evaluation significantly changed the behavior of the occupants of the facility due to the hope and expectation that

The **reactive effect**, where building users **change** their behavior because of the POE in progress, is **difficult to prevent**.

working conditions would improve. Building users and client organizations may change their behavior and responses because of the presence of POE team members. In fact, just the anticipation of a study may prompt a variety of changes in the building to be evaluated.

This reactive effect is difficult to prevent or document since most of the anticipatory activity occurs before the on-site evaluation begins. The building users will most likely be informed of the project early, in compliance with the National Research Act discussed later in this chapter. However, careful observation, the use of photography, and some interviewing regarding recent changes in the building can be useful.

One way to ameliorate the reactive effect is to meet the building occupants and explain the nature of the POE. Emphasis should be placed on POE as building- and not personnel-focused data collection and analysis; the goal should be defined as the improvement of buildings and not people.

Guarding against the reactive effect in a POE and reducing occupants' anxieties are traditional concerns. In practice, the authors have experienced the opposite reaction. The building users are generally enthusiastic about a forthcoming POE and are more than willing to spend considerable time talking about the building being evaluated. The anxieties caused by a POE may be more of a problem for the evaluators than the building occupants.

2. MONITORING AND MANAGING DATA-COLLECTION PROCEDURES

Quality control in data collection is essential. To accomplish this, procedures should be established to assure the consistency and usefulness of the data to be collected.

Quality control in data collection assures the **consistency** and **usefulness** of the data to be collected.

Inter-observer Reliability

Inter-observer reliability is based on carefully orchestrated comparisons of observations made by different observers of the same scene, that is, a check of whether the same phenomena are being recorded in the same way by different people.

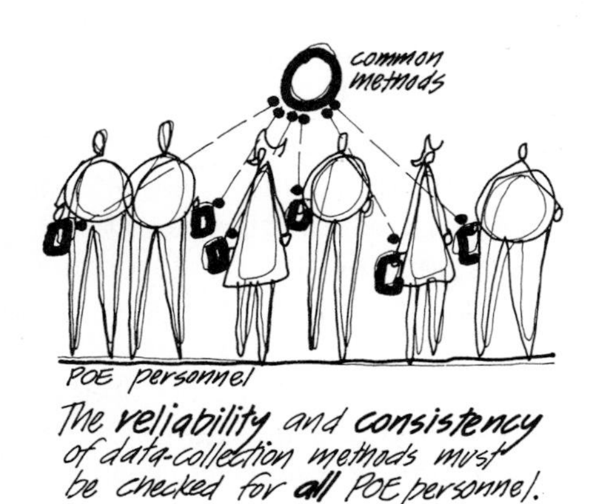

The reliability and consistency of data-collection methods must be checked for all POE personnel.

This procedure is generally used during practice runs and only occasionally during data collection. Answers to several questions are critical: are the evaluators holding sound-level meters in the same position and orientation within rooms? Are the instrument readings taken on the same "scale" setting? Is a conversation across an open office recorded in the same way as workers talking in a group?

Clear and definitive procedures should be specified, disseminated, practiced, and monitored. Since procedures can never be planned completely, on-the-spot solutions need to be used by the evaluators in the field during the POE.

Pretesting of Data-collection Instruments

Pretesting of data-collection instruments such as questionnaires and interview questions is commonly carried out in the form of a pilot study, using representatives of the type of respondents who will be sampled in the actual POE. The purpose of pretesting is to assure that the data collection will yield valid and useful data, and thus it helps avoid wasted POE efforts. Questions may be interpreted in various ways by different building users. The words *environment* or *programming* may have different meanings to different people. Asking building users about the closeness of an amenity may provide more information about the respondents' perception of closeness than factual information about the distance itself. While both pieces of information may be useful, they refer to different concerns. Pretesting, therefore, is standard practice in POEs. A few practice runs not only will help in assuring consistency of data collection but also will familiarize the evaluators with the process and increase the efficiency of the evaluation.

Both questionnaires and interview question sets should be pretested before they are actually used.

Data from pretests and practice runs are used for quality control and are subsequently discarded. Data should be carefully monitored at the beginning of actual data collection; even some of these may be discarded if problems are discovered. The scheduling of indicative or investigative POEs may not allow this luxury, but at the diagnostic level of POE, where data are collected over longer periods, this step is required.

Practical Guidelines

This section deals with tools, shortcuts, and other how-to advice for conducting a POE in the field. All the items discussed here are not employed in every POE, however.

PROTECTION OF HUMAN SUBJECTS

In 1974 the National Research Act became law and created the National Commission for the Protection of Human Subjects in Biomedical and Behav-

*For all POEs, especially investigative and diagnostic, building users must be protected from **risk** and invasion of **privacy**. This is especially important in the study of **behavioral** elements.*

ioral Research (Office of Protection from Research Risks 1983). A few years later, the Department of Health and Human Services adopted it as its policy in this area. Although officially these regulations only apply to work connected with this agency, many universities and research institutions have adopted similar rules and approval committees for all research involving human subjects.

Few of the technical and functional evaluations carried out in the course of a POE require approval. Indicative POEs should generally be exempt from such reviews because of their more cursory nature. However, investigative and diagnostic POEs that involve behavioral elements of performance will need to ensure that human subjects' rights, such as privacy, are respected and protected. To that end, many institutions, such as schools and hospitals, require a formal review and approval process involving the signing of consent forms by subjects prior to the onset of a POE.

POEs are virtually harmless in terms of risk to the people who participate. High on the "protected" human-subjects list, for instance, are prisoners, patients, handicapped persons, children, pregnant women, and fetuses. High-risk procedures, for example, may include certain types of medical research such as testing new drugs, using clinical devices, and administering electrical shock. These examples and a review of the POE literature indicate that POE efforts are benign in terms of their effect on human subjects. In fact, many projects have benefited their human subjects. Thus, it would be ironic if a review of human-subjects protection resulted in discouraging POE projects with potential user benefits.

The review process is relatively simple. It includes a description of subjects, research procedures, and risks involved in data collection. It describes how subjects are safeguarded, identifies the confidentiality of data, and delineates the means for informing subjects about the proposed research procedures. The name of a neutral party is supplied, usually a member of the approving organization, to be consulted if subjects have questions concerning the procedures that are used in a POE.

Questionnaires are not subject to review if the questions do not require personal evaluation or ask for private information, and, therefore, surveys of a factual nature do not pose any problem.

There are also certain types of research that are exempt from human-subjects protection review. These include:

- Research conducted in established or commonly tested educational settings, such as the evaluation of instructional strategies or the comparison of classroom management methods
- Educational tests
- Existing records if the subjects cannot be identified
- Survey or observational research, *except* if the subject can be identified and the subject's responses could place him or her at the risk of criminal liability or loss of financial standing or employability

If there are any questions about the proper methods and procedures to be used in POE research, universities and research institutions usually have a Human Subjects' Protection Review Committee that may be contacted for advice. Local representatives of the Department of Health and Human Services can also be consulted.

Photography is an inexpensive data-collection tool that provides a number of benefits.

PHOTOGRAPHY

Whatever level of POE is employed, photography will play a key role. A photographic record of certain building conditions and activities will allow for reexamination after the on-site work is completed.

Data Collection

Photography can be used directly to collect data or to complement other methods. Photography is particularly useful for collecting behavorial data. Analyzing photo documentation and field data together, after the on-site visits, allows time for recording additional detail and occupant information, particularly when many occupants are involved in a number of activities. This technique has been used to study such environments as urban plazas, school playgrounds, and prison yards. Variables recorded through photography include the number of persons, group size, interpersonal distance, subjects' orientation, posture, activity, time, and weather conditions.

Photography can also be used as a primary method for the collection of data on functional elements of performance, such as handicapped accessibility, storage, flexibility, equipment usage, and furnishings.

In using photography to collect data on functional and behavioral elements of performance, particular care must be taken to identify each photographic frame on-site, by location, orientation, date and time, descriptive data, and notes on the subject matter. After processing, each image should be coded and checked to match the on-site sequence. Photographic data can be recorded, aggregated, analyzed, and checked.

The camera is also a good way to document the POE process for later publication and presentation.

POE Process Documentation

Photographs that document the POE process are useful for publication and presentation purposes. The activities in the building, the on-site POE process, and highlights of successful and unsuccessful elements of building performance can be shown.

Color slides have been found by the authors to be most versatile for data collection on both functional and behavioral elements of performance. Slide images can be enlarged considerably for detailed analysis, and slides are easily stored, catalogued, and presented.

A high-quality 35mm camera with interchangeable lenses is excellent for most POE photography, but the authors have found that a compact, reliable, automatic 35mm camera works just as well, particularly for the documentation of user behavior. This type of camera eliminates the focusing, adjustments, and attention required by more sophisticated cameras.

Many types of technical and functional performance elements can be evaluated in the evenings or on weekends.

AFTER-HOURS DATA COLLECTION

Evenings and weekends are suitable times for the evaluation of technical and many functional performance elements for all three types of POEs. With no interruptions and continuous working time, an impressive amount of data can be collected. For example, a typical POE scenario in an office situation would be to begin after the occupants have left the building, about 5 or 6 P.M. By working systematically and continuously through the evening, elements such as the interior finishes, acoustics, HVAC, and lighting can be evaluated. With a large POE team, a number of functional factors such as the amount and type of storage, anthropometrics, and support equipment can also be covered. In addition to the above, access on weekends allows the examination of the exterior walls and roof surfaces.

After two or three hours of POE work, the team should take a short coffee break and discuss progress. The evening's work might end around midnight. The role of experienced principal evaluators at the beginning of such "marathon" data-collection sessions will be to manage and monitor the quality of the collected data. If problems occur, much valuable time and data may be lost. Later in the POE the principals can participate in the routine data-collection process themselves.

Cassette **tape recorders** can be used to collect data, conduct debriefings, and summarize findings.

TAPE RECORDER

The microcassette tape recorder has become an important tool in conducting the on-site portion of a POE. It has been used by the authors in collecting interview data, debriefing of POE team members after meetings, and summarizing data-collection segments as they are completed.

In an evaluation of a research laboratory building, three methods were employed to collect data on the use of utilities (water, air, and so on), equipment, and glassware. On a reduced building plan the utilities and location of equipment were noted. Photography was used to document elevations of the labs, and a microcassette recorder was used to describe the equipment and its location, to note the type and location of glassware, and to keep track of informal comments. All three categories of data—utilities, equipment, and glassware—were coded for the location, date, and time of data collection. When moving through a facility, the recorder becomes a kind of efficient notepad, rather than a dictating machine.

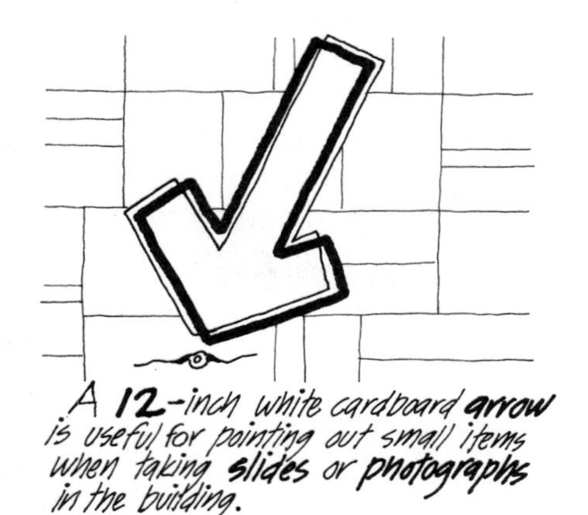

A **12**-inch white cardboard **arrow** is useful for pointing out small items when taking **slides** or **photographs** in the building.

OTHER USEFUL AIDS

The "Big Arrow," spontaneously created during an on-site POE, has become a standard and often-used aid. Made of stiff white illustration board and edged in bright orange marker, this 1-foot-long arrow is used in photography to call attention to specific, often small items. For example, a long, narrow crack, one-sixteenth of an inch wide, might not easily be identified in an ordinary photo or slide. When the Big Arrow is included, the situation becomes immediately clear. Primarily employed in the evaluation of technical performance elements, the arrow has been effective in showing indentation, spalling, and cracks. In evaluating functional performance elements it has been used to point out specific types of storage, display, and equipment. For close-up photography in which detailed phenomena may be scrutinized, a scale drawn on the arrow provides a reference measure to show the relative size of the identified objects.

Other useful items in the POE kit include masking, duct, and filament tape, a pocket knife, various screwdrivers, markers of various sizes and colors, a powerful magnifying glass, and a flashlight.

The building maintenance staff is an excellent source of information about the building.

SOURCES OF INFORMATION

The most undervalued source of information in a POE may be the building's maintenance staff, whose members are usually intimately familiar with the building, its history, the organization occupying it, and the users. Often, different maintenance personnel are concerned with different aspects of building operations and maintenance. Janitorial staff members are most familiar with the interior finishes (particularly flooring) and their durability and cleanability, and repair personnel are knowledgeable about the ongoing maintenance of such items as plumbing, electricity, locks, and doors, as well as being responsible for small repairs and decorating. Both janitorial and repair personnel may also have insights into patterns of use in building areas such as those that receive heavy use (wear and dirt), have inadequate utility service (electrical problems, extension cords), and the like.

A maintenance superintendent is usually cognizant of large building problems, and he can provide the POE team with repair and renovation records of the building over time.

Expensive POE measurement equipment must be stored in a safe place when not in use.

SECURITY

Because instruments used during the evaluation process are usually sophisticated and expensive pieces of equipment, they should be safeguarded against theft or vandalism. A secure place in the facility should be available, either lockable or under watch. Security problems can arise when automatic data-recording instruments, such as recording thermometers and sound-level meters, are left unattended.

Additional care should be taken not to breach the security of the facility being evaluated. During evening hours and weekends, the evaluation team may be responsible for the building and should allow no one but the team admission to the building. The evaluation team members should be aware of their responsibility, and the facility should be left in the same condition it was in at the beginning of the data-collection process.

Finally, the day-to-day activities in the facility must not be disrupted by the POE team. In evaluating a chemical research laboratory that houses dangerous substances, for example, changes in light levels and temperature caused by the POE team's activities may affect the functions carried out in the building.

3. ANALYZING DATA

Purpose of Data Analysis

Although data analysis occurs while the on-site phase of the POE is conducted, the analysis will have been anticipated in terms of methodology and resources in the research-planning step of the POE planning phase (see chapter 5). The effort necessary in analyzing data is often underestimated, even for an indicative POE. When more sophisticated analysis techniques are used, the research-planning phase becomes more important.

The purpose of analyzing data is to identify response patterns or in other ways differentiate among the findings of a POE. The analysis step in the POE process tries to make sense of the data in terms of the research questions asked at the outset of the POE. Occasionally data can help predict the per-

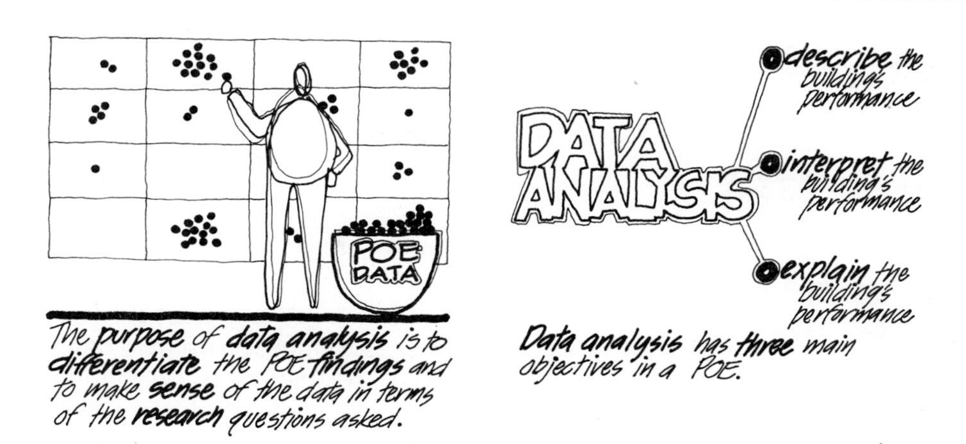

*The purpose of **data analysis** is to **differentiate** the POE **findings** and to make **sense** of the data in terms of the **research** questions asked.*

***Data analysis** has **three** main objectives in a POE.*

formance of a building or building type so that the outcomes can be used to feedforward information into databases and, subsequently, into design guidance and criteria literature.

In its broadest sense, data analysis has three objectives for the investigator, client, and professional audience for POE findings:

1. *To describe* the performance of the building and its elements
2. *To interpret* this performance and judge its merit
3. *To explain* this performance

Another step in the process, *recommending* future action, takes place in the application phase of the POE process, after considerable discussion with the client organization (see chapter 7).

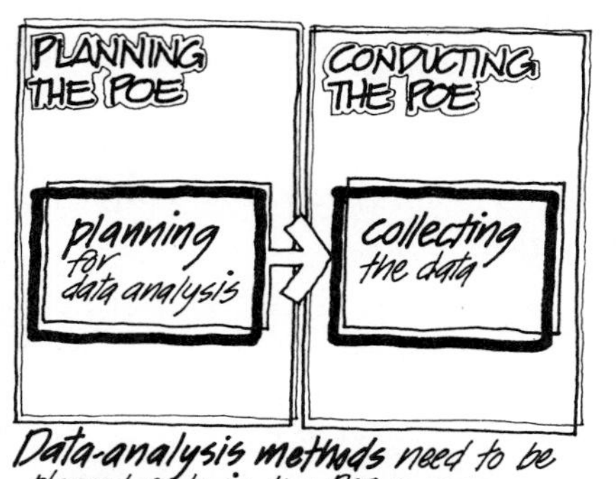

***Data-analysis methods** need to be planned early in the POE process because they affect the way data are actually **collected**.*

The analysis of data expected from POEs must be planned at the outset of the project since it affects the data-collection methods to be used and the level of effort of the POE. Analyses at the indicative level of POE typically yield qualitative and quantitative indicators of successes or failures in building performance. These may be in the form of simple counts of the number of times a given item is mentioned or observed as a result of a walk-through evaluation, or analysis may take the form of interpretations of documentary evidence and photographs of, for example, building abuse or overuse. In an indicative POE, analyses reflect the building occupants' gross perceptions of and aggregate responses to aspects of building performance. As such they suggest certain problems that may exist, but they are in no way conclusive.

At the investigative level, simple, descriptive, statistical techniques may be used, and therefore the analysis of data will be based more on fact than on the intuition of the evaluator or building occupants. This level of POE typically still deals with one case or building, and therefore the analysis yields information that is primarily applicable to the project context and is not generalizable.

At the diagnostic level of effort of POE, cross-comparisons of building types are made, requiring representative samples, and therefore more elaborate statistical analyses are carried out.

Types of Data Analysis

indicative POE — data — analysis framework

investigative POE

diagnostic POE

The sophistication of the data-analysis techniques vary with the type of POE being done.

Depending on the level of effort and type of POE that is conducted, different levels of sophistication in data analysis are applied. Methods of analysis may range from simple tabulations, frequency counts, and standard deviations to cross-tabulations and sophisticated analysis techniques such as factor analysis and analysis of variance (see glossary). Choosing the appropriate method depends on the purpose and scope of expected evaluation results and their ultimate use, as well as the budget and schedule.

For most simple POEs that are based on singular case studies, descriptive statistics will suffice since "quick-and-dirty" findings are sought and the time and expense of sophisticated data analyses are thus not warranted.

The application of statistics is a complex topic on which many textbooks have been written. For most of the simpler descriptive types of statistics, there are statistical analysis methods or packaged programs that instruct the user in setting up and formating data to be run in analysis routines on personal computers.

There are several basic data-analysis techniques, ranging from common sense to sophisticated statistical techniques.

For the majority of POE projects, data analysis involves three common-sense steps.

COMMON-SENSE DATA ANALYSIS

For most indicative and investigative POEs, data analysis will be a straightforward process primarily involving interpretation, judgment, and explanation of the results. In the less frequent diagnostic POEs and those that include behavioral performance elements, data analysis will involve more sophisticated methodologies.

A complete analysis will include the on-site performance measures, performance criteria, the degree of success or failure of the building element being evaluated and an explanation of the level of success. An indicative POE, however, would include only a brief analysis, primarily addressing the success or failure of the building elements being evaluated. For many aspects of a POE the use of the following analytical techniques is feasible, particularly in the technical and functional categories of performance.

Success / Failure

A number of building attributes evaluated at all three levels of effort may have a single success / failure as a result. Many basic building elements such as the

elements of performance	success	failure
code compliance	✓	
temperature		✓
lighting		✓
fire safety	✓	
door widths	✓	

The evaluation and analysis of many building attributes lead to simple success or failure judgments about those elements' performance.

elements of performance — success … failure

esthetics

functional efficiency

user satisfaction

The evaluation and analysis of some building attributes lead to judgments about degrees of success or failure.

exterior wall and roof and code-related attributes such as structure, ventilation, temperature, light levels, and the many aspects of fire safety very simply either work or do not. Functional performance elements, such as whether doors are sufficiently wide to allow passage for replacement of equipment when needed or whether utilities are adequate to accommodate certain types of changing uses, may also fall into this category.

Rating Scales

The use of rating scales as analytical techniques allows the description and presentation of different degrees of performance that are applicable to a great number of building-performance situations. The study of buildings in use in chapter 10 used 95 percent, 85 percent, and 75 percent ratings to judge the performance of many elements and is based conceptually on a previous study of carpets done at the Wharton School of Management (Parks 1966). In this rating scale, 95 percent represents excellent performance, 85 percent a level of performance needing some maintenance, and 75 percent represents a level requiring "outside contracting." Why these elevated criteria? In most buildings even very small dysfunctions are easily visible and annoying: the worn carpet, the light bulb that flickers, the buzzing fluorescent light, the cold office, and the skylight that occasionally leaks. These deficiencies may not be very significant, but their presence even on a small scale has a noticeable effect.

SEMISOPHISTICATED DATA ANALYSIS

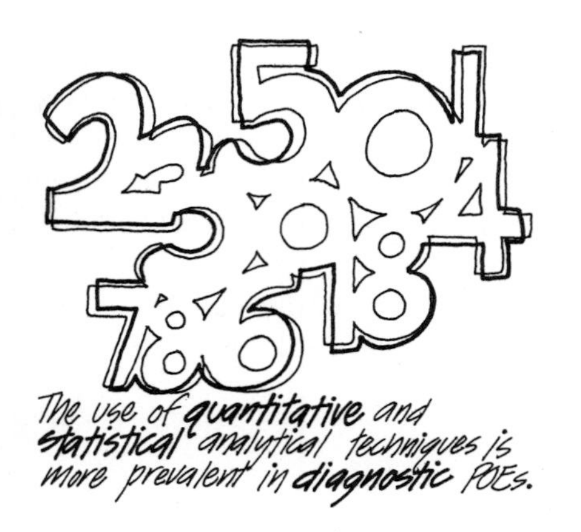
The use of **quantitative** and **statistical** analytical techniques is more prevalent in **diagnostic** POEs.

The following analytical techniques are particularly appropriate for diagnostic POEs and some aspects of investigative POEs.

Averages and Measures of Central Tendency

The analysis of data is the search for patterns in data that may help explain relationships or summarize data among the variables being examined. A basic category of methods for analyzing data is the measure of central tendency—summary values that reveal where the most frequent values occur or which are representative because they are in the middle of the data. Measures of central tendencies provide a single statistic that represents the data. The average or mean is the most well known measure of central tendency. Averages and measures of central tendency are useful, but they obscure the actual frequencies and qualitative differences that may be experienced in buildings, and, therefore, analytical methods in recent POEs have become more sophisticated.

Percentages

Data analysis in survey research and interviews often computes percentages to explain and interpret the aggregate data. Percentages are worthwhile in describing and presenting data in POEs where easily discernible differences can be found, for example, percentage of usage of certain rooms can easily be computed from space-utilization schedules.

Measures of Variability

Measures of variance and standard deviations are techniques that provide a picture of the concentration of data, for instance, a mean with a relatively high standard deviation indicates that there is relatively little agreement among respondents. On the other hand, a low standard deviation indicates strong agreement concerning the item being rated.

SOPHISTICATED STATISTICAL DATA ANALYSIS

The following types of analysis require knowledge of statistical techniques. These techniques are most frequently employed in diagnostic POEs, specifically in the analysis of behavioral elements of performance. The techniques are simple and apply inferential statistics typically taught in introductory statistics courses. There are a number of concerns such as sampling an unwieldy number of variables or difficult field conditions that may require the hiring of statistical consultants.

In buildings of more than 25,000 square feet or great complexity, it may not be possible to evaluate every area. A procedure for sampling should be developed that represents the various conditions encountered in the facility. Such conditions may include similar rooms representing various directional orientations, window configurations, user types, or administrators. Even if sampling is part of an overall evaluation strategy, a cursory, indicative POE should be made of every space in the facility.

Comparing Two Groups

Differences may emerge in comparing two groups of data by using either measures of central tendency or percentages. Are the differences between groups significant? If the evaluation were repeated, are the differences still found to be reliable? While the averages may differ, for example in the test scores of two groups drawn from the same sample, data from each group may overlap. A "t-test" is a measure of the differences between two group's means that are not just chance. The high levels of significance usually used as criteria in this test provide evidence that observed differences are caused by the variable being examined.

Simple Analysis of Variance

Analysis of variance is used when comparing data from three or more groups from the same population. It is conceptually similar to the comparison of two groups in that differences in observed mean scores among the groups are examined to determine whether at least one group score is significantly different from the others.

Chi-square Analysis

The tests for differences between two groups and the analysis of variance are applied to continuous or interval data such as test scores. How should categorical or nominal data such as male/female, Republican/Democrat, smoker/nonsmoker be tested? The Chi-square procedure examines the differences between actual and expected frequencies between groups and whether the differences between these measures occur by chance or are statistically significant. An example of categorical data that may be analyzed is the smoking/nonsmoking habits of Republicans and Democrats.

Correlation Analysis

Unlike the above analyses, which provide measures of validity for the differences between groups, correlation analysis provides measures of association among variables. For instance, a technique that has been applied to a number of POEs is Spearman's rank-order coefficient, which measures correlation in ranking. One can rank each student's closeness to the teacher's desk (the closest is ranked number 1) and also rank the student's test scores in order to discover any pattern of correlation.

Typical Tasks in Analyzing Data

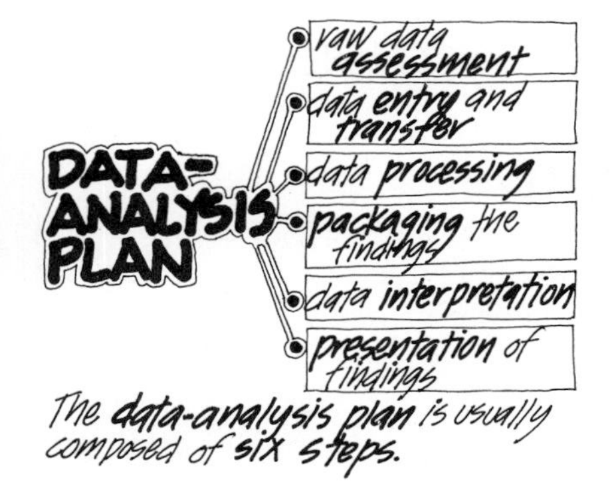

The data-analysis plan is usually composed of six steps.

The degree of sophistication in data analysis largely depends on what the client organization wants to gain from a POE. The client's objectives are translated into the type of data that must be collected and the analysis that must be performed (see "Research Planning" in chapter 5). The data-analysis plan will be initiated when the so-called raw data have been collected.

TASK 1: ASSEMBLING RAW DATA

Raw data result from the data-collecting step in the POE process. They may be in the form of questionnaires, interview notes, time logs of activities, videotapes, frequency counts of occurrences or events, or various types of documentation.

TASK 2: DATA ENTRY AND TRANSFER

Task 2 consists of entering data that may be on survey forms, observation log sheets, maps, or interview tapes, for example, into a format that permits quantitative treatment of those data. In many cases, this requires the establishment of a coding system according to predetermined data categories. In some cases unexpected data are yielded from the data collection, particularly when open-ended questions are involved, so that the categories of data and the appropriate coding can be assigned only after the raw data are assembled.

Where standardized data categories and measures are concerned, it is often possible to use precoded data forms called scanning sheets that permit the respondent to enter answers by marking the sheet with a pencil in such a way that an automatic scanner can "read" the data forms.

TASK 3: DATA PROCESSING

In simple tabulations of frequencies or standard deviations, the computation involved in the data analysis can be carried out by hand or with a small pocket calculator. More involved analyses may require statistical software for a personal computer or, for large projects, the use of a mainframe computer. If that is the case, the cost and time involved in data processing should be carefully considered at the outset of the POE.

TASK 4: PACKAGING AND COMMUNICATING FINDINGS

When data have been processed and are analyzed, they become findings or results of the POE. By organizing the findings appropriately, the evaluator can make interpretations, discover patterns or trends, and, later, begin to formulate conclusions based on what has been discovered. Processed data require communication formats that facilitate quick access, legibility, and clarity of presentation involving such visual methods as bar graphs, computer graphics, and diagrams.

TASK 5: DATA INTERPRETATION

Based on solid information, the analysis and interpretation of processed data require a good deal of creativity and integration on the part of the evaluator. Data in themselves may not say very much, but when seen in the context of a client organization and specific building, they may indeed provide insights or pointers that would not otherwise be self-evident. At this point in the POE process, the multifaceted approach to data collection begins to pay off; various sources of information concerning building performance will have been tapped, and the task of correlating and integrating the resulting data is required to obtain a holistic picture of the successes or failures of the subject building.

*The way the POE findings are **presented** should relate back to the POE **objectives** and **expectations** of the client.*

TASK 6: COMPLETING DATA ANALYSIS

This is the final part of conducting the POE. Findings are presented and organized according to a predetermined hierarchy of issues, such as building scale from the site scale down to design details, or according to a set of priorities established beforehand by the client organization. For example, the POE findings may stress health and safety issues and not deal at all with functional or psychological aspects of building performance. In a comprehensive POE all these aspects of performance will be addressed and may be presented according to funding priorities that may exist. In other words, there is no set formula for the presentation of POE findings; the original objectives of the client organization will dictate presentation formats.

Data-analysis Personnel

In situations where simple descriptive statistics are required in the process of analyzing data, the evaluator may choose to carry out the work in the interests of quick turnaround and reducing cost. In more involved statistical analyses, especially where correlational techniques are involved, the POE team may need the services of a specialist in statistics.

*The **level** at which data analysis is to occur must be **matched** by the **analysis skills** of the POE **personnel**.*

Pitfalls in Data Analysis

While data collection and analysis are planned activities, the reality of conducting the POE sometimes differs from the plan.

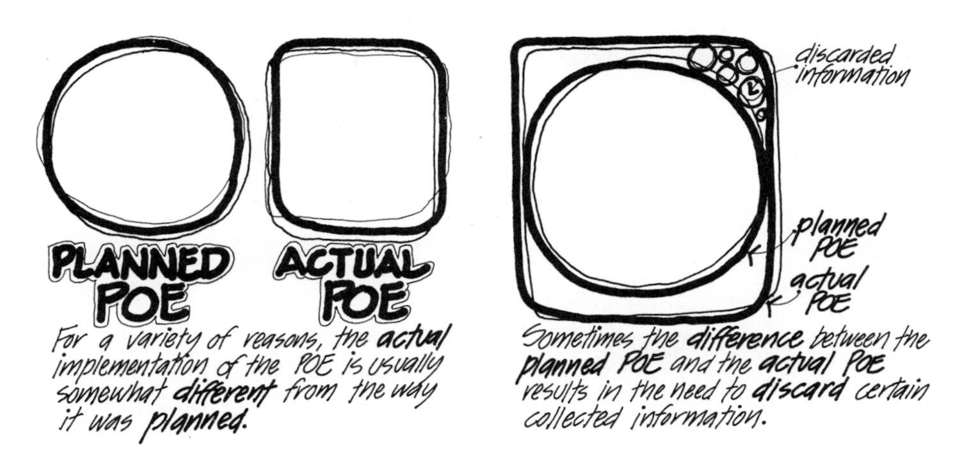

The many reasons usually relate to difficulties experienced with on-site data collection. Nonrepresentative samples may be discovered, invalid responses may have been made on survey questionnaire forms, or data may have been tampered with. This means that at times data must be discarded and new data collected.

In a POE conducted in the military, the authors found that certain survey responses were invalid because a company commander had dictated the "right" answers to two hundred respondents. In the same type of authoritarian and hierarchical organizational structure, interviewing or testing respondents in the presence of supervisors proved to be of no value since the respondents did not dare to bring up controversial subjects.

In some cases data collection is hampered by seasonal conditions, inaccessible respondents or building areas, or a client organization that fears the unauthorized release of confidential or controversial data. Other causes of possible flaws in collected data can be equipment failure during data collection or errors made by the POE team. For example, in certain POEs respondents are given response material to react to, such as slides. Due to photographic distortions, inclusion of irrelevant scenery, uneven weather conditions, and other flaws, slides can be misleading or misrepresent the subject to be evaluated. Along similar lines of thought, biased groups of subjects may have been sampled due to abnormal conditions such as weather or special events that may prevent certain groups of subjects from being present in the environment being evaluated.

Conducting the POE has its foundations in the planning that went before it. Though, happily, there are discoveries to be made during the on-site phase, a POE must be rigorously executed and depends on formal and systematic inquiry.

7. Applying the POE

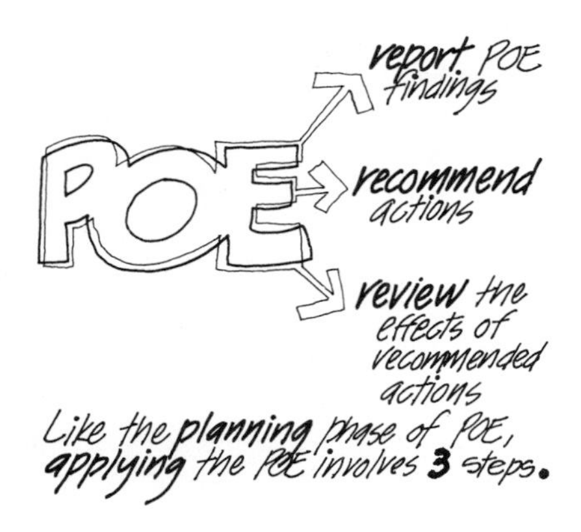

The major activities in applying the POE are reporting the findings, recommending and planning actions, and reviewing the effects of those actions on the client's facilities and operations. These three implementation activities fulfill the original goals of the POE process.

REPORTING FINDINGS

- solve short-term problems
- resolve major problems
- input findings to information systems

*There are **short-term**, **medium-term** and **long-term** objectives in reporting the findings of POEs.*

The findings of POEs can be reported in several ways depending on the client organization and the purpose or use that the POE findings will serve. The findings of POEs are primarily oriented toward four objectives:

1. Identifying problems and evaluating successful and unsuccessful performance in existing facilities. This type of POE may be used in rating alternative buildings for purchase and occupancy by an organization. Findings would include such items as inadequate space or ventilation.
2. Achieving short-term problem solving by identifying and resolving minor building and space problems. Such findings typically imply remodeling, minor construction, changes in space utilization, furnishings and equipment, and possible staff relocation. These actions, requiring limited resources, are often accomplished through the use of in-house staff. Findings for short-term objectives are typically in the realm of improving maintenance and operations of the facility and making changes in relatively short-lived building elements and furnishings, such as carpets.
3. Resolving major environmental and space problems in existing facilities. The findings obtained in the POE may result in changes in the structure of both the organization and the existing facilities, including major construction and the addition of buildings or parts of buildings. Solving these problems can take from three to five years, but the solutions

should meet organizational requirements for decades. To accomplish such medium-term objectives, POE findings are focused on areas such as access for the handicapped, the capacity and flexibility of utility systems to serve changing telecommunications and computer needs in the future, and the capacity of the building to accommodate growth. Also considered are replacement needs for major building elements, such as lighting and storage capacity, as well as changes in behavioral factors.

4. Affecting the long-term operation of the client organization. POE findings can be fed into information systems. The focus is on the development of design criteria to be used in databases—the purpose of which is to improve the quality of new construction as well as replacement of major systems. These findings are based on a variety of performance assessments including:
   - *Building configurations:* net and gross area ratios, energy utilization, user orientation, circulation, safety
   - *Location of activities:* capacity and density, privacy, territoriality, proximity, usage, sociometrics, workflow, communications, supervision, flexibility, location and servicing of equipment
   - *Technical systems:* durability, maintenance and repair, appearance, function, flexibility, building-systems coordination
   - *User response:* scale and image of the building, impact of amenities, comfort, access and circulation, visual stimulation, order, esthetic considerations

Report and presentation formats and mechanisms will vary significantly based on the objectives of the POE and the level of effort involved.

1. Reporting Indicative POEs

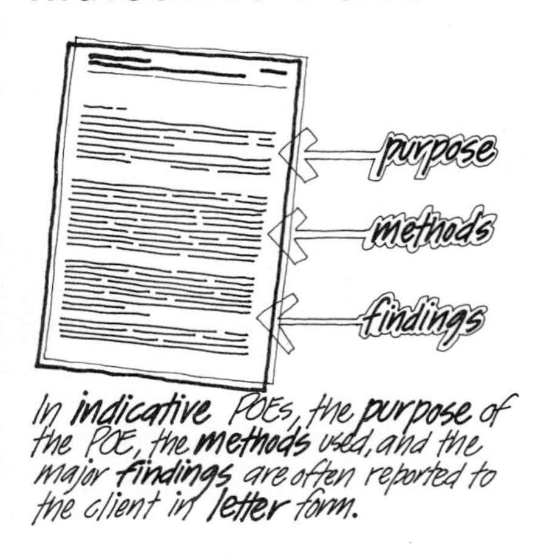

In **indicative** POEs, the **purpose** of the POE, the **methods** used, and the major **findings** are often reported to the client in **letter** form.

Indicative POEs usually focus on problem identification and short-term objectives. Findings are customarily reported in the form of a three-to-eight-page "letter report" highlighting the significant findings of the walk-through type of investigation. A typical letter report contains the following:

PURPOSE

The objectives and focus of the POE are briefly stated.

METHODS

A brief outline of data-collection procedures is given, including the specific methods used.

FINDINGS

Significant findings should be reported and prioritized. Safety and security issues, for instance, may be the most important priorities, the adequacy of space may be next, and so on. For each significant finding, a recommendation for action should be included, either calling for further study or offering specific guidance.

2. Reporting Investigative POEs

The findings of investigative POEs are usually presented in the form of reports that contain both the evaluation criteria and performance measures for each element studied.

An investigative POE that focuses on medium-term objectives requires a short report, from fifteen to fifty pages long, depending on the scope of the POE and the depth of investigation. Explicit measures of building performance and criteria are provided for each element of performance measured. An evaluation of each attribute is made. Recommendations for action are often included. An investigative POE report includes annotated plans and may use photographs or drawings to emphasize and clarify findings.

An investigative POE is often used by the client organization to document shortcomings of existing facilities as evidence in funding applications. An investigative POE report can be supplemented by a videotape summary or slide presentation that graphically illustrates the POE findings. Typically, the client organization will receive a briefing at the conclusion of the POE in which these media can be used effectively. In turn, the client organization can use media presentations to strengthen the case for funding renovations, additions, or new building construction.

3. Reporting Diagnostic POEs

Diagnostic POEs are often documented in detailed, technical, and often voluminous reports. It is advisable to include an executive summary at the beginning of such reports. The format is usually that of a research-based report including the following parts:

- Executive summary of findings
- Introduction
- Methodology
- Data analysis
- Findings
- Conclusions
- Appendix
- Bibliography

The voluminous reports from diagnostic POEs are applicable for use in databases and future design projects.

The appendix may contain a description of the building, statement of goals, outline of the administrative structure of the client organization, surveys of occupants, annotated list of data-gathering instruments such as survey forms and interview schedules, and a summary of data collected. There may be excerpts from space standards, accident reports, and space-utilization schedules. Similar state-of-the-art facilities in other locations may be documented in the form of floor plans and commentaries by the owners or occupants.

The results of diagnostic POEs, because of their comprehensiveness and complexity, are most applicable for feedforward into databases and for the design of subsequent buildings of the same type.

RECOMMENDING ACTIONS

The completed POE should always ***recommend actions*** *that need to be taken by the* ***client organization.***

A critical part of a completed POE is to recommend actions to be taken by the client organization. The nature of these actions will depend on the initial purpose of the POE, but basically there are two types of recommendations. Recommendations may address a hierarchy of possible implications in terms of modifications to or innovations in policies, procedures, and techniques employed by the client organization. Recommendations may relate to different time horizons for their implementation, organized in hierarchical fashion, i.e.: short-term, medium-term, and long-term.

Policy-related Actions

A POE can suggest many policy-related actions. For example, policy-related actions may be recommended to change a client organization's basic use of its building stock. In one case, it may be prudent to abandon a production facility that may still be reasonably efficient and functional from a production point of view but in the foreseeable future will be made obsolete by new work processes and technologies. Tax considerations may prompt a policy that accelerates the program for replacing the facility. In another case, the policy recommendation may be to improve facilities-management procedures in order to maximize space utilization. This might involve reorganization of production shifts, relocation of certain parts of the organization's activities, or elimination of some activities where appropriate.

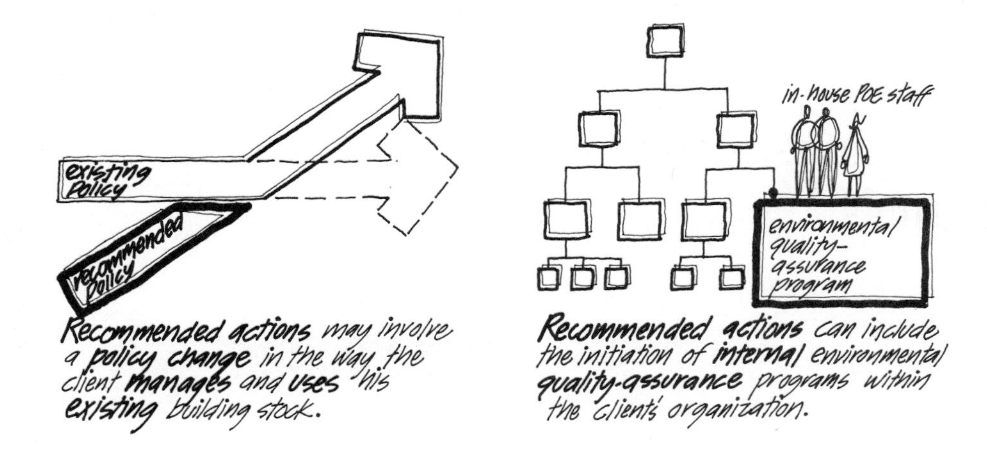

Recommended actions *may involve a* ***policy change*** *in the way the client* ***manages*** *and* ***uses*** *his* ***existing*** *building stock.*

Recommended actions *can include the initiation of* ***internal*** *environmental* ***quality-assurance*** *programs within the client's organization.*

Quality assurance is a topic of increasing importance at a time when greater accountability is demanded of corporations and large institutions. Quality assurance can be considered an internal "auditing" mechanism that can be carried out routinely by in-house personnel with POE capability.

Other policy-related action pertains to the growing trend toward "networking," the sharing of information through information databases and central clearinghouses. Again, POE can make important contributions to those aspects of the performance evaluation research framework (see chapter 3).

Litigation is a concern of those who build, own, and operate facilities. POE can provide a credible source of information on the building performance in legal cases (see appendix A). Obviously, policy-related action in this area would aim at avoiding litigation or, if unavoidable, at keeping penalties to a minimum.

Procedure-related Actions

RECOMMENDED PROCEDURES

- space utilization
- security
- communication
- access to high-level staff
- evaluation of new technology
- safety
- code compliance

Procedure-related recommended actions can affect ***specific client activities*** *within his or her facility.*

Recommendations pertaining to procedures followed by the client organization can again vary widely and are typically addressed to different levels of management, such as executives, middle management, or support staff. For example, the client organization's management structure and style greatly influence space utilization, security measures, communication among organizational units, access to certain levels of personnel, and other activities. In a strongly heirarchical organization, such as the military, procedures will be more formal and followed more rigidly than in the private sector. In the first case they may be documented in technical manuals or other guideline literature, which may not always be the case in the private sector.

As "bottom-line" thinking becomes pervasive and as the concept of life-cycle costing becomes accepted, facilities are increasingly being regarded as strategic assets. POE will prove to be a useful tool for facilities managers by permitting continuous monitoring of quality and utilization of facilities.

Procedure-related actions stemming from POEs can also be applied to the evaluation of new materials and technologies as they are introduced into the building industry. As is the case with many products, the safety of new materials and technologies must be tested and potential hazards eliminated. Evaluation procedures involve actual "consumers" of buildings, that is, representative samples of occupant groups.

Technique-related Actions

Recommended actions concerning techniques used in the client organization may pertain to such issues as operation and maintenance techniques

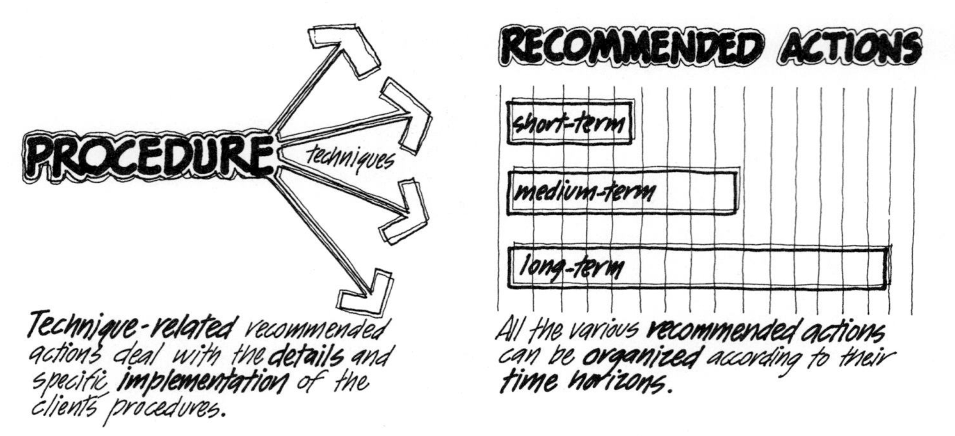

Technique-related recommended actions deal with the ***details*** *and specific* ***implementation*** *of the client's procedures.*

All the various ***recommended actions*** *can be* ***organized*** *according to their* ***time horizons.***

that would maximize the beneficial use of a facility. For example, this might mean using cleaning materials to enhance the appearance and reduce the potential deterioration of building finishes.

As "smart" buildings become common, it will be feasible and useful to employ integrated building-monitoring systems that permit the continuous collection of data on building performance and utilization and the continuous adjustment of the building in response to these data.

Recommended actions pertaining to different time horizons can be grouped into actions with short-, medium-, and long-term implications, as indicated in the following examples.

SHORT-TERM ACTIONS

Immediate, fix-it–type actions are those that will solve problems without major cost or construction efforts. In many cases, organizational changes may take care of the problems without even having to change the facility. For example, parts of the organization may have underused space while others may be overcrowded. A shift in space use and subsequent relocation of departments may resolve this kind of problem.

Actions involving modest modifications in the facility, such as repainting, relocating office partitions, and exchanging or pooling furnishings and equipment, are part of this short-term response.

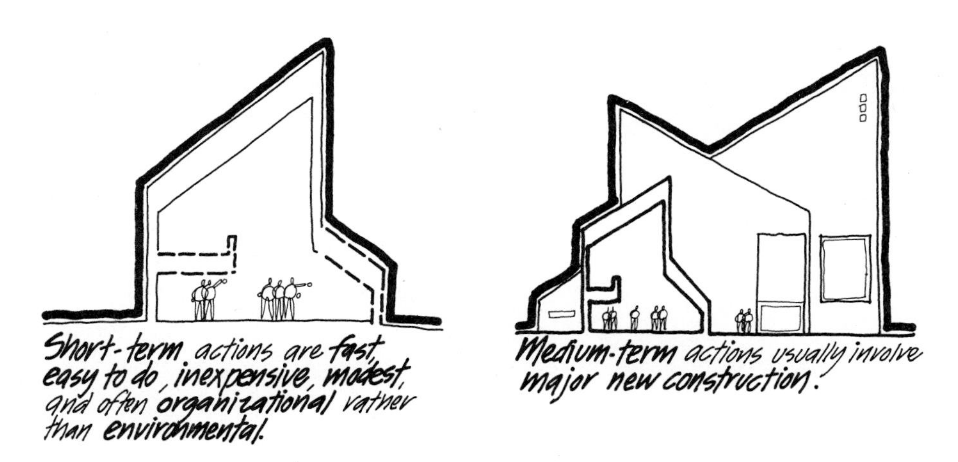

MEDIUM-TERM ACTIONS

These actions lead to major new construction, such as demolishing part of the facility and replacing it with a new building or addition, or abandoning an existing facility and planning and designing a new one. This would take, on average, from three to five years from the initiation of the plans to move-in, and would require major capital outlay.

LONG-TERM ACTIONS

Long-term actions imply diagnostic research in support of the development of design guides and criteria for future, similar facility types. This option is in line with the notion that the levels of effort of POEs are evolutionary in nature, —the more in-depth information is expected, the greater the POE level of effort will have to be. Therefore, the recommendations for action in some cases will suggest the implementation of the next higher level POE, by progressing from indicative to investigative and finally diagnostic POEs.

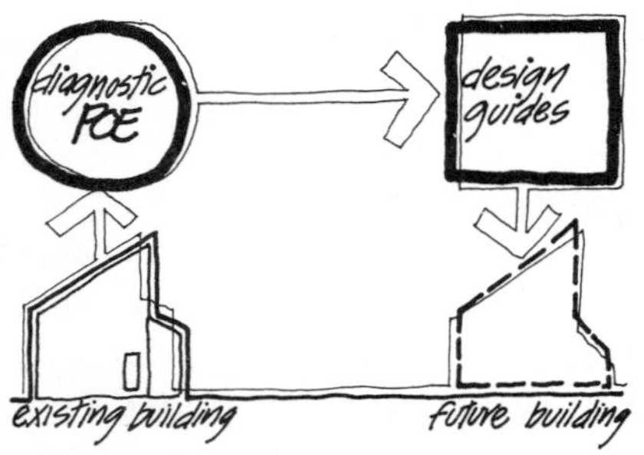

Long-term actions often relate to diagnostic research to develop design guides and criteria for future similar facilities.

Sometimes recommended action suggests the implementation of the next higher level of POE.

It is usually a good idea to urge the client to action while the POE is still fresh in everyones mind.

In general, recommendations for action are made through close coordination of the POE evaluator team and the representatives of the client organization. Often it is advantageous to urge action while "the iron is hot," that is, while members of the client organization who have participated in a POE are still involved with the critical issues and eager to see them resolved.

In many cases the POE report and the recommendations for long-term action can be used effectively to obtain increased organizational and funding support by providing concrete evidence and documentation concerning facility usage and by serving as justification for required funding and action.

Programming and Design

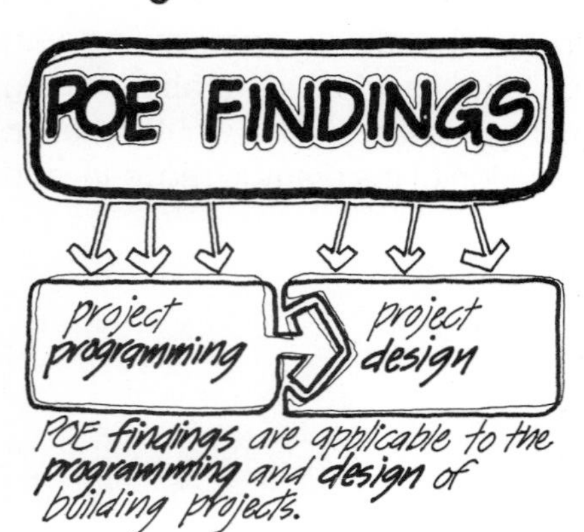

POE findings are applicable to the programming and design of building projects.

POE findings are directly applicable to the programming and design phases in the building process. For example, in a laboratory situation, specific POE findings may indicate that substantially more desk-top writing surface and less laboratory work surface are required, that additional short-term and long-term storage is needed, and that certain larger pieces of equipment are not well accommodated. A diagnostic POE would provide programmers with specific criteria for needed desk surfaces, bench areas, and storage capacities. While indicative and investigative POEs could indicate some of these problems, they would not enable adequate quantification of these needs.

The programmers can integrate these types of findings, along with other data, and then provide specific figures for space needs, in addition to diagrammatic layouts of areas and services needing close proximity. In the design process the materials, connections, joints, and overall form are determined, based upon the program and technical POE findings, such as the durability of specific laboratory bench materials and the resistance of surfaces to chemicals.

REVIEWING OUTCOMES

The review of outcomes ensures that recommended actions are implemented and achieve the intended results.

This step in the POE process ensures that recommendations and actions that were the results of a POE are followed, put to use, and achieve the intended results. This is essentially an evaluation of the evaluation process and the subsequent steps to implement its outcomes. Such a review of the outcomes should not occur immediately after completion of a POE; rather, a transitional period should be allowed so that the reactive effects of POE-caused changes are ameliorated. This is particularly important in the case of long-term consequences of POEs, which may take many years to have any noticeable and measurable effect. Communication and contractual links must be maintained between the evaluators and the client organization in order to facilitate review of long-term outcomes of POEs.

For government or other large organizations, POE activities may be carried out by in-house staff on a routine, cyclical basis as long as effective feedback mechanisms are devised. Such mechanisms should permit the regular updating of technical manuals, design guides, and databases through feedback from field offices in charge of facility planning and management. In this context, integrated, computerized databases assume increasingly significant roles as conditions in the building context change over time.

Part 3

Post-Occupancy Evaluation Case Studies

8. Indicative POEs of Four Senior Centers in Albuquerque, New Mexico

PROJECT SYNOPSIS

The objectives of this project were to evaluate four senior centers, to apply the lessons learned from their successes and failures to a generic program for future senior centers, and to use this program in the design of a new senior center. The development of performance requirements for a generic senior center program involved three major phases:

1. A literature assessment was conducted on the topic of senior centers, focusing on goals, activities, and environmental requirements for typical space categories, as well as building and site design considerations.
2. Walk-through POEs were conducted on four senior centers that had been built since 1976.
3. Programming workshops with user-agency representatives combined their expertise on senior centers with the outcomes of the first two phases of the project.

Major POE findings included: problems with unprotected, confusing or dark entries (figs. 8-1 and 8-2); irregularly shaped, wasteful spaces; windowless spaces; inadequate kitchen facilities; some lobby and office spaces with circuitous circulation patterns; and outdoor activity areas that were basically unused. Positive performance aspects included: the lobby areas' and pool

Figure 8-1. Unprotected entry at Highland Senior Center.

Figure 8-2. Dark and confusing entry at Palo Duro Senior Center.

Figure 8-3. Socializing in Highland Senior Center lobby.

Figure 8-4. Dancing in Highland Senior Center multipurpose room.

Figure 8-5. Flexible food service line.

rooms' excellent potential for socializing; security of gift shops; dance floors in multipurpose rooms; and a very flexible arrangement for food-service lines (figs. 8-3–8-5).

The project result was a generic program for future senior centers in Albuquerque, New Mexico, which was subsequently modified and used in the design of the Far Northeast Heights (Bear Canyon) Senior Citizens' Center, also in Albuquerque. It formed the basis of a POE of this center after it was in use for one year, and it also became the basis for a second center that is now in the planning stage.

The entire project was carried out over a six-week period. Walk-through POEs required only one-half day per senior center, including preparation and setup, data gathering, and debriefing. The entire time required for the project was estimated at about 120 man-hours. The evaluator team consisted of an architect and an architectural researcher who were supported by secretarial staff. An overview of project data is contained in table 8-1.

1. Project Name:
   Indicative POEs of Four Senior Centers in Albuquerque, New Mexico

2. POE Level of Effort:
   Indicative, walk-through POE

3. Purpose/Use:
   Feedforward into generic program and design

Table 8-1. Project overview.

4. Building Type:
   Senior centers (4)

5. Building Size:
   Approximately 12,000 sq. ft. each

6. Building Age:
   3–10 years (at time of evaluation)

7. Building Location(s):
   Albuquerque, New Mexico

8. Client:
   City of Albuquerque, New Mexico

9. Important Factors Analyzed:
   Technical
   Leaks, breakage, maintenance
   Functional
   Adequacy of space and environment; access for the disabled
   Behavioral
   Control of access; psychological comfort (noise, smoke)

10. Methods Used:
    Data collection
    Literature review, interviews, survey, still photography, observation
    Data analysis
    Simple frequency tabulations

11. Project Personnel:
    1 architectural researcher; 1 architect; volunteer committee; staff support

12. Project Duration/Man-hours:
    6 weeks/120 hours

13. Project Cost:
    $8,000

14. Major Lessons Learned:
    State-of-the-art design criteria do not necessarily apply to all contexts; changing demands of facility type require flexibility (such as senior day-care)

15. Benefits to Clients:
    Generic program adaptable to future projects

Table 8-1. (Continued)

1. PLANNING THE POE

1.1 Reconnaissance and Feasibility

The project was initiated by the City of Albuquerque Office of Senior Affairs. The city interviewed more than twenty competing architects from New Mexico who were interested in conducting background research, including POEs, and in designing a senior center in Albuquerque. The contract was awarded to an architect in collaboration with an architectural research consultant. A special add-on contract was drafted for consulting services that went beyond standard architectural services.

An indicative POE was chosen because of the limited time, manpower, and other resources available for the project. In planning the POE, a major requirement was to obtain and use the POE results within only a few weeks. It was determined that three to four hours at each POE site would be sufficient. An evaluation plan was devised using three types of methods outlined in the project synopsis: literature search and assessment; archival research; and walk-through evaluations including interviews, on-site inspection, and photographic documentation. The project initiation phase required one week out of the six that were used for the entire project.

Basic data on the existing four senior centers in Albuquerque were obtained, including floor plans, listings of program activities, and changing functional uses over time, such as the introduction of meal programs for the elderly and adult day-care programs within senior centers.

This phase involved archival research and the review of documents provided by the client organization. In some cases, building plans were not available or were not up-to-date, and as-built drawing had to be produced based on actual measurements of the buildings involved. A comparative analysis was conducted on the amount of space devoted to various activities common to all four senior centers.

1.2. Resource Planning

Senior center directors were informed of the project and the input to be sought from staff and clients. The City of Albuquerque Office of Senior Affairs was instrumental in identifying liaison persons at each of the four senior centers who would assist in the coordination of site visits, interviews, and discussion sessions with senior center clients. In addition, an Ad Hoc Committee on Senior Center Planning was organized, consisting of knowledgeable persons, all volunteers, who had extensive experience in senior center programs and knew the issues pertaining to the use of senior center facilities.

The overall project was managed by the architectural firm of Barker/Bol & Associates, which was the prime contractor for the POE and design portions of the work, as well as the liaison to the client representative for the Office of Senior Affairs. The subcontractor, an architectural researcher, was responsible for the content and procedure of the POE work described in this case study. The project team consisted of a representative from the architectural firm and the architectural research consultant who collaborated throughout the project. This permitted continuity of information flow from project initiation through the design phases.

Operational planning of the project involved an estimate of the number and type of persons in each of the senior centers who would be involved in the information-gathering phase. Given the very short time frame, only a few representatives of senior center staff and clients were interviewed. Diagrams showing senior center staff organization and the various program activities were used to identify prospective interviewees. Equipment used in the POEs was limited to tape measures and 35mm cameras. Clipboards were required for note-taking and sketching during the walk-through evaluations, and small cassette tape recorders were used for on-site dictation of notes during the walk-through evaluation.

A schedule was drawn up for the intended four walk-through POEs, which were carried out within one week.

1.3. Research Planning

Evaluation criteria for senior centers were limited to an assessment of published state-of-the-art literature that focused on two topic areas: community and service centers in general and senior centers as they have evolved in the United States, with additional attention being paid to day-care programs for elderly persons who are fragile. That concept, first introduced in the United Kingdom, is now becoming an important component of senior center programs in the United States in an attempt to integrate as much as possible the fragile elderly with those who are able-bodied.

A literature review started with computerized, on-line literature searches of relevant databases including Psychological Abstracts, the National Technical Information Service, and the Smithsonian Science Information Exchange (see appendix C).

Literature sources were screened for project applicability, particularly with regard to issues and concepts in the design of senior centers. Subsequently, selected sources were reviewed to establish performance criteria for the POEs of the senior centers. The nature of information found in the literature ranged from overall goals and objectives for the health and psychological well-being of the elderly population in the United States to highly specific facility- and space-related data and recommendations. Subjects included successful spatial relationships and circulation patterns, successful room configurations and environmental provisions, and hardware specifications. Most useful in this regard were the publications by Jordan (1975, 1978), whose overall goals and considerations for the programming and design of senior centers (see table 8-2) became the baseline against which the performance of senior centers in Albuquerque was measured. The information that resulted from the literature assessment was reviewed by a panel of experienced senior center coordinators and users who evaluated it for applicability to the context of the elderly in the Southwest and specifically Albuquerque.

The elements of the performance concept in the building process (see fig. 3-3) that link user behavior with performance elements of the physical environment were used to organize the findings from the literature. The users' requirements were stated in terms of human goals and objectives as they pertain to the elderly population in general. In addition, the implications for the design of senior centers were outlined in terms of spatial and environmental considerations.

A translation of these goals and objectives into specific spatial and architectural provisions in senior centers was made with the direct input of experi-

1. Increase opportunities for individual choice
2. Encourage personal independence
3. Reinforce the individual level of competency
4. Compensate for sensory and perceptual changes
5. Recognize some decrease in physical mobility
6. Improve comprehension and orientation
7. Encourage social interaction
8. Stimulate participation
9. Provide individual privacy
10. Reduce distractions and conflicts
11. Provide a safe environment
12. Improve the public image of the elderly
13. Provide functional and accessible outdoor space
14. Plan for growth and change

Table 8-2. Examples of goals for senior centers.

enced senior center coordinators and users. The resulting performance criteria became the standards used in the subsequent POEs of the four senior centers. These evaluations permitted feedforward into the programming stage of the project. Such criteria were established at different levels of scale and concern regarding the built environment: the entire building site, the overall building performance, and the performance of generic spaces such as dining, meeting, or office spaces, examples of which are provided later in this case study.

Data-collection methods for the walk-through POEs involved two techniques: recording comments from the occupants of the senior centers and recording and documenting visual physical evidence of the use patterns of the facilities.

A list of fifteen evaluative questions (table 8-3) was given to senior center directors with a request to have these questions answered before the site visit and walk-through POEs would actually take place. The purpose of the evaluative questions was to help the client organization and its representatives focus on essential building performance-related issues. Responses to the questions helped in orienting senior center staff to issues addressed by the evaluation team, and, implicitly, to prepare the participants in the evaluation for the on-site interviews.

Interviews followed the schedule and sequence of the evaluation questions pertaining to a given senior center. In addition, issues that had surfaced in the written responses to the generic questions and subsequently had been highlighted during the walk-through of the respective senior centers were discussed. Interviews were open-ended and served to elicit detailed information about issues identified by senior center staff.

A second set of methods was used as part of the process of recording the physical measures of performance of senior centers. The methods selected for collecting these data were observation, physical measurement, and still photography, all of which were used during the walk-through part of the evaluations.

SENIOR CENTER BUILDING EVALUATION

We would like to know how well your Senior Center performs for all those who occupy it. Both successes and failures are considered, including features that affect occupant health, safety, efficient functioning, and psychological well-being. Your answers will help improve the design of future senior centers.

NAME: *POSITION:* *CENTER:*

Please identify successes and failures in the building by responding to the following broad information categories and by referring to documented evidence or specific building areas wherever possible.

1. Adequacy of Overall Design Concept
2. Adequacy of Site Design
3. Adequacy of Health/Safety Provisions
4. Adequacy of Security Provisions
5. Attractiveness of Exterior Appearance
6. Attractiveness of Interior Appearance
7. Adequacy of Activity Spaces
8. Adequacy of Spatial Relationships
9. Adequacy of Circulation Areas, for example, lobby, hallways, stairs
10. Adequacy of Heating/Cooling and Ventilation
11. Adequacy of Lighting and Acoustics
12. Adequacy of Plumbing/Electricity
13. Adequacy of Surface Materials, for example, floors, walls, ceilings
14. Underused or Overcrowded Spaces
15. Other, please specify (such as needed facilities currently lacking)

Table 8-3. Building-evaluation questions.

Planning for data analysis included simple tabulations of the data, specifically counts of the occurrence of certain phenomena or the expression of priorities and the frequencies of issues that surfaced during the evaluation and interview sessions with selected senior center staff and users. A sophisticated, time-consuming data analysis was not considered appropriate for this indicative POE.

Preliminary criteria on senior center evaluation were limited to those resulting from the assessment of the state-of-the-art literature. They consisted of generalized goals and objectives for senior centers, which in turn became the basis for a generic set of questions. These were directed toward senior center staff and discussion sessions with senior center users that usually followed the walk-throughs. The interviews were to be conducted prior to, during, and after the walk-throughs.

The directors of the senior centers were advised of the proposed dates for the walk-through, and they in turn informed staff and senior center users of the upcoming visits. Throughout the entire POE process, the Ad Hoc Committee on Senior Center Planning was asked to review the results of the project at various stages.

2. CONDUCTING THE POE

2.1. Initiating the On-site Data-collection Process

Prior to the date of the walk-through POEs, the set of fifteen generic questions concerning building performance was sent to the directors of the senior centers in order to help them understand the POE process and prepare for the planned interviews. The evaluation team brought tape recorders, notepads, and clipboards for the interview sessions and still cameras for the walk-throughs.

2.2. Monitoring and Managing Data-collection Procedures

The entire procedure of interviewing senior center directors and selected staff, the walk-through of the facilities, and the subsequent discussion sessions with senior center users took approximately three hours at each center. The sequence of data collection started with an interview of the center director, who provided an overview of the positive and negative characteristics of the respective senior center. This was followed by the actual walk-through of the facility, during which the items mentioned in the interview and briefing were pointed out in the respective center areas by the senior center director and staff. During the walk-through, the two-member evaluation team conducted brief interviews with staff and selected users, taking notes and recording observations. Occasionally, evaluators made sketches of the use of spaces or noted special, unusual occurrences. In most cases, the evaluators worked independently of each other. Later in the data analysis, their notes would be compared. The walk-throughs were followed by group discussions with around twenty senior center users who had indicated interest in the project and had specific concerns they wished to have addressed. These concerns ranged from management to the planning and design of senior center facilities.

In these POEs, key persons to be interviewed among the staff of the senior centers included the administrative, janitorial, and kitchen staff, as well as staff members running the many programs and classes, i.e.: programs ranging from medical checkups and food service to handicrafts and other classes. In addition, persons in charge of the regularly occurring social events, such as dances and entertainment, card games, and billiards, were contacted at this time.

The walk-through POEs concluded with a debriefing of the senior center director and key staff members prior to the departure of the evaluation team.

2.3. Analyzing Data

Data-analysis methods used in the four indicative POEs were simple, in keeping with the expenditure of time and resources appropriate to that level of effort. Since the members of the evaluation team had familiarized themselves with the state-of-the-art of senior center planning and design prior to conducting the walk-through POEs, they were quickly able to identify both positive and negative features of the senior centers being evaluated—even though these centers differed greatly in the quality of design and amenities they provided.

Recordings of interviews, notes of discussion sessions, and direct observations by the evaluation team were transcribed and analyzed. Data were compared with the evaluation criteria and organized according to a hierarchy of different building scales at senior centers, such as the site scale, building scale, and room scale. The frequency of specific issues mentioned by respondents was noted.

3. APPLYING THE POE

3.1. Reporting Findings

The findings of the four senior center indicative POEs were organized and reported according to scale of environment—from considerations concerning the overall site and the entire facility to those concerning generic space categories, such as food-preparation, administrative, and meeting spaces, through to detailed aspects of building performance such as lighting, surface characteristics, and color. Because the focus of this book is on the process of POE, not the detailed findings of case studies, only selected representative findings at each level are included. (For full details, see Preiser and Pugh 1986b.)

Reporting the findings of POEs involves not only the accurate and precise compilation of data, but also appropriate communication formats.

Palo Duro Senior Center was built in 1974, the first senior center in Albuquerque, followed by the 7th Street Barelas Center (1978), the North Valley Center (1980), and the Highland Center (1980). On a comparative basis, the performance of the oldest of the senior centers was considered least satisfactory. Major problems at Palo Duro Senior Center included irregularly shaped, windowless, and wasteful spaces that were difficult to furnish (fig. 8-6), overcrowded and unsafe kitchen facilities (fig. 8-7), and very poor lobby and office spaces (fig. 8-8).

In the United States, most centers have experienced the need for expansion within relatively few years, primarily because programs have changed and interest in and demand for senior activities have increased. Thus the need exists to provide for flexibility, growth, and expansion in senior centers.

Following are sample summaries of evaluation criteria and findings from the walk-through evaluations for lobby and senior day-care areas. Lobbies were found to be the prime locations for meeting, socializing, watching friends, waiting for rides, playing games, and keeping up with the activities at the centers (fig. 8-3). As hubs of activity, lobbies were generally found to be appropriately placed, adjacent to the main entrance and the office/administration areas. Users and staff of one center that had several entrances complained about difficulties in keeping track of center clients and in finding friends. The best lobbies provided for ease of circulation, clear views to the entrance and parking lot, and a variety of seating arrangements allowing groups of various sizes to congregate for conversation, games, or waiting.

Because of differing physical abilities, different kinds of seats were found desirable, soft and firm, large and small. Users generally complained about chairs that lacked armrests to assist them in getting up or sturdy backs to use as supports when standing.

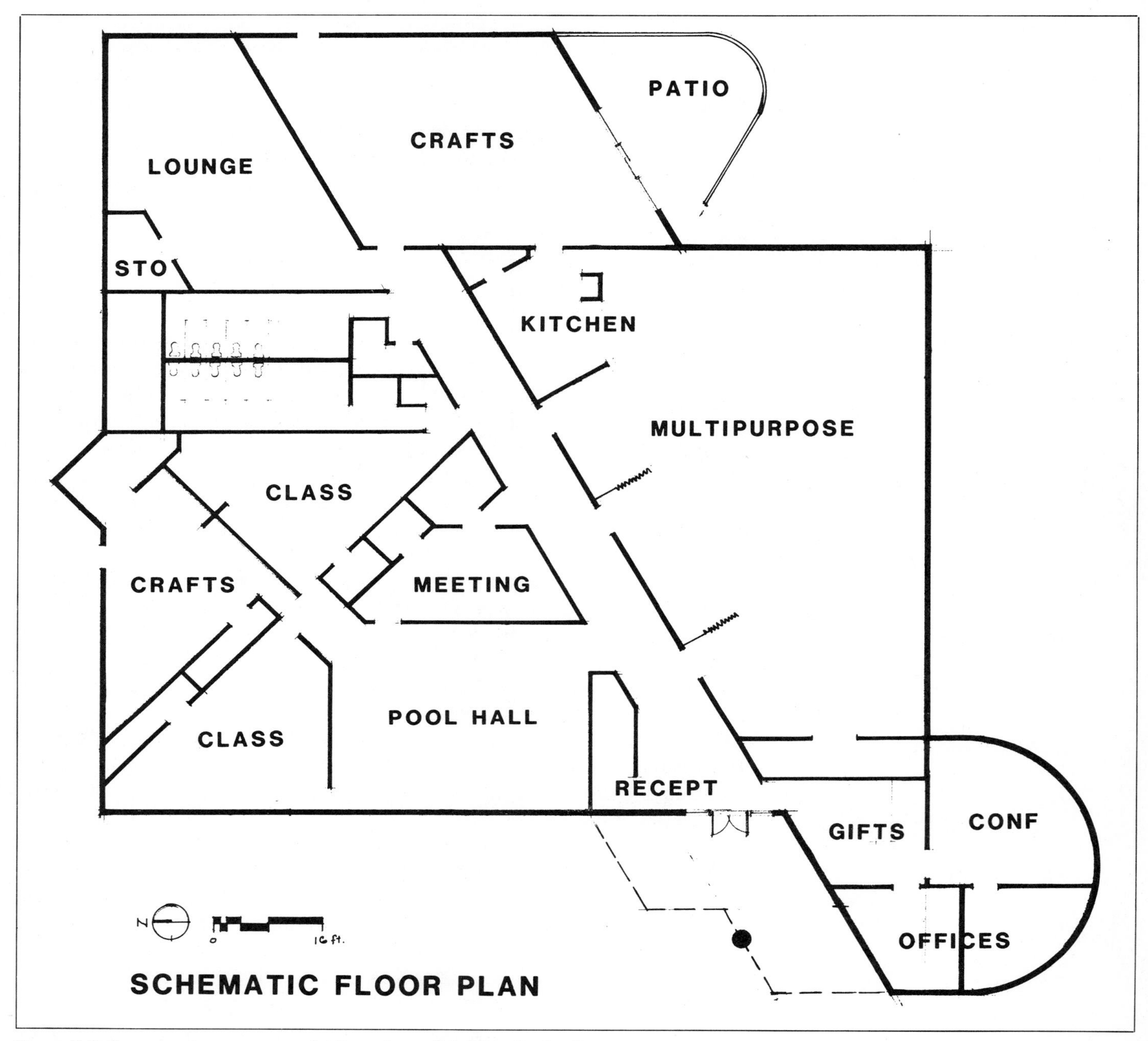

Figure 8-6. Example of an unsuccessful floor plan—Palo Duro Senior Center.

Senior center clients appreciated well-lit, cheery places. One dark lobby was being retrofitted with brighter lighting at the time the POE took place. Because many users spent the greater part of their day in the lobby, natural lighting and views to the outdoors were considered to be highly desirable. Bulletin boards were universally used as communication centers, and ample lighting was desirable in these areas as well. Because of the mix of smokers and nonsmokers, adequate ventilation was required. The comparisons between evaluation criteria (table 8-4) and actual building performance constituted the act of evaluation.

Figure 8-7. Overcrowded kitchen with unsafe, protruding oven door.

Figure 8-8. Long, narrow lobby discourages social interaction.

REQUIREMENTS: ENTRY/LOBBY/RECEPTION AREA

1. Occupants
   Director, consultant, staff, center users
2. Activities/Time
   Reception, administrative, counseling, registration

3a. Health/Safety and Security Requirements
   Main entrance sheltered from rain/snow and well lit
   Visual supervision of building entrance by receptionist
   Safekeeping of money

3b. Functional Requirements
   Reception counter height without bending
   Clear storage for direction finding
   Waiting space
   Office 1 (secretary, receptionist)
   Office 2 (director)
   Office 3 (seniors' own office)

3c. Psychological Requirements
   Main entrance image: visible to community, well lit, and good signage
   Limit access to one (some) offices for privacy, for example, director's for confidential conversations, in addition to group seating

Table 8-4. Program requirements for entry/lobby/reception area.

(continued)

4. Occupant-Equipment Requirements
   Electrically operated entrance doors recommended
   Public address system with good acoustic quality
   Well-lighted bulletin board
5. Ambient Environment
   Daylighting in entry/lobby
6. Locational Requirements
   Adjacent to entrance: offices and lobby
7. Special Requirements
   Entrance at grade for access for the disabled
   Plan for future expansion of office space for outreach programs and volunteer counseling
   Avoid heavy doors
8. Area
   Entry/lobby, circulation area: 475 sq. ft.
   3 offices: 250 sq. ft. each

Table 8-4. (Continued)

3.2. Recommending Actions

The generic program for designing senior centers in Albuquerque represents the synthesis of the preceding two phases with the results of programming workshops that were conducted according to the model outlined in *Problem Seeking* by Peña et al. (1977). Participants in these workshops were knowledgeable individuals who were actively involved in the management, administration, and programs of senior centers, as well as representatives of actual user groups. Two workshops were conducted over the period of four weeks. The programming workshops were evolutionary: in the first one, generic program requirements and overall design concepts were presented to the group, and detailed program requirements and performance criteria (table 8-5) were reviewed for applicability to the Albuquerque context. In the second workshop, overall spatial relationships and priorities concerning site and orientation were discussed. The development of prototypical concepts as a basis for schematic design was then achieved by manipulating actual space sizes and categories two-dimensionally. This was accomplished with color-coded pieces of cardboard fitted into spatial relationships that were discussed by the group and recorded with a Polaroid camera (fig. 8-9). At the end of the second workshop, four prototypical concepts had been identified and prioritized by the workshop participants. The most successful conceptual program diagram (fig. 8-10) was used in the design of the Far Northeast Heights Senior Citizens' Center (fig. 8-11). The program requirements referred to here were the basis for a POE that was conducted after occupancy of the new facility (Kirkpatrick et al. 1986), showing its superiority in virtually all aspects of building performance.

REQUIREMENTS: SENIOR DAY CARE

1. Occupants
   Frail, dependent elderly; 3 permanent staff; occasional volunteers. As many as 15 to 30 elderly will use the facility from time to time.
2. Activities/Time
   Daytime operating hours. Activities include: preparing and eating snacks; light exercising; conversing and watching television; doing arts and crafts; performing RSVP activities such as stuffing envelopes, folding, and stapling; occasional sleeping and napping; some singing and dancing.

3a. Health/Safety and Security Requirements
Provide for total accessibility for the disabled. Avoid sharp edges, rough and slippery surfaces.

3b. Functional Requirements
Provide a multipurpose room that allows for exercise, dance, slide presentations, movies, and music. A food-preparation and eating area, a separate arts-and-crafts area, a sleeping area that can be darkened and where privacy can be provided by use of a screen, a separate office for the staff, a location for personal items such as coats and hats, and a place to sign in and out.

3c. Psychological Requirements
Keep the space bright, cheery, alive, homey, and deinstitutionalized. Provide a place for watching the world go by. Allow the occupants to see outside the center and to observe other activities on a daily, regular basis. Provide a secure environment that allows for as much independent action as possible.

4. Occupant–Equipment Requirements
   Provide storage for individual items. The kitchen should have a small freezer/refrigerator, a small range, a place to wash dishes, and a place to eat. The arts-and-crafts area should have a small sink, a storage area, and a work table. Provide for the possibility of a piano and a stereo system for music.
5. Ambient Environment
   Provide natural lighting and sunshine. Provide for individually zoned heating and air conditioning. Provide soft, safe surfaces.
6. Locational Requirements
   The day-care area should have a separate, covered entry with van pickup and dropoff point. It should be located near the senior center's dining and multipurpose area. Staff offices in the day-care center should be buffered and separated from the offices in the senior center.
7. Special Requirements
   Provide optimum accessibility for the disabled.
8. Area
   Approximately 2,000 sq. ft.

Table 8-5. Program requirements for senior day-care.

Figure 8-9. Programming workshop number 2.

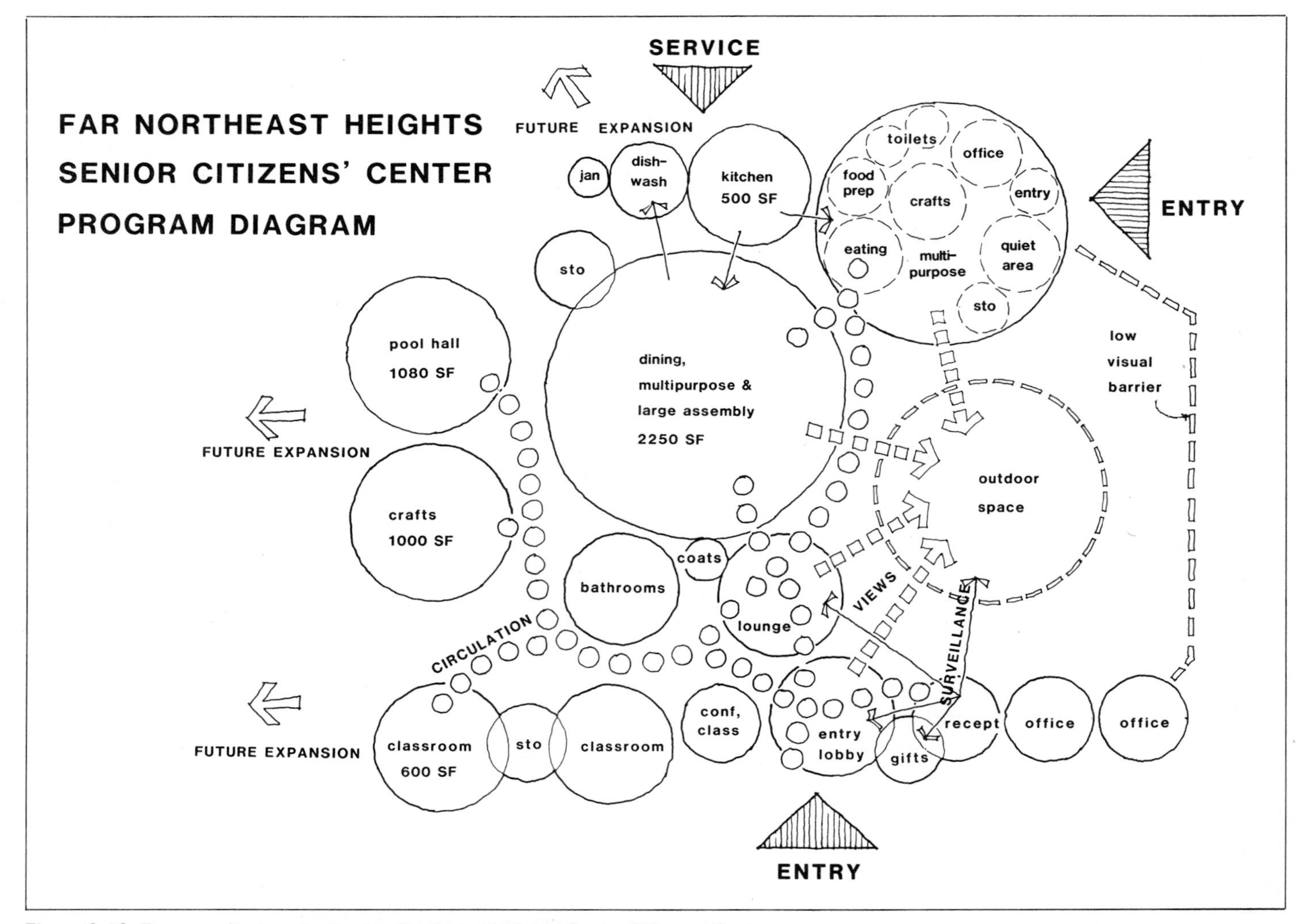

Figure 8-10. Program diagram for the Far Northeast Heights Senior Citizens' Center.

3.3. Reviewing Outcomes

The three-phase approach described in this chapter—literature assessment, the indicative POE, and programming workshops—was considered by all involved to be very successful, particularly since local expertise, including representatives of the eventual users, was made part of the process. The participants had a vested interest in seeing the project through to a successful conclusion. They helped verify basic programming assumptions by scrutinizing the performance criteria that had been distilled from the literature on senior center design. Finally, they expressed their great satisfaction at being involved in the entire process.

The City of Albuquerque Office of Senior Affairs provided the following feedback at the conclusion of the project:

> The project undertaken to evaluate the four senior centers operated by the Office of Senior Affairs in Albuquerque has been most beneficial to us.

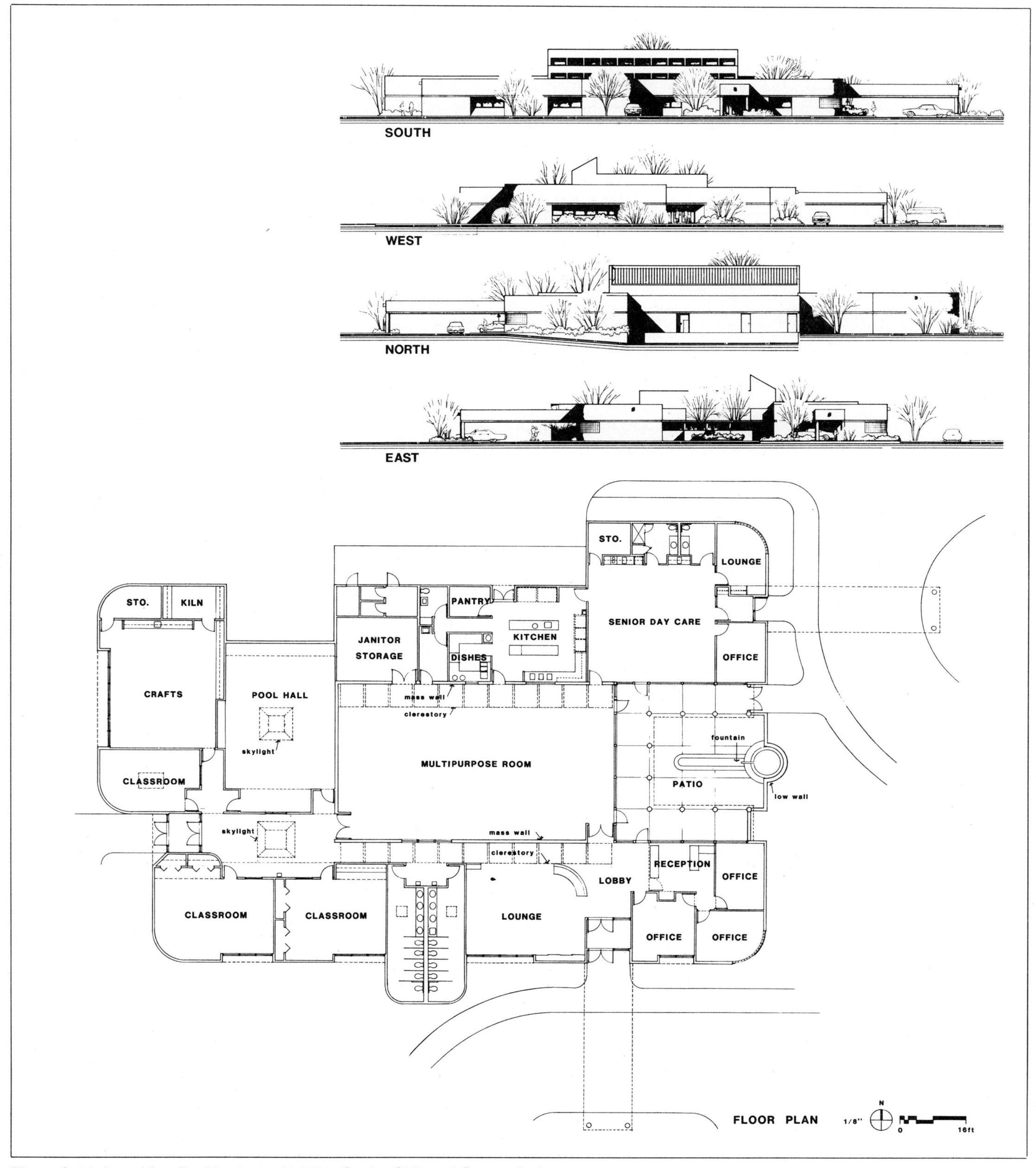

Figure 8-11. Resulting Far Northeast Heights Senior Citizens' Center design.

> The information that resulted from the project, particularly the post-occupancy evaluation document, was used extensively in the development of the generic program for future senior center development.
>
> The approach used included a questionnaire, interviews with clients and staff, the actual walk-throughs and the programming workshops. It was invaluable in making design decisions that were easily adaptable to the site. This process gave the staff and the clients an opportunity to express themselves and a sense of pride in the outcome.
>
> We are quite pleased with the final document and have recommended it highly to others in the field of aging.
>
> Ginger Grossetete, Operations Manager
> Office of Senior Affairs
> Albuquerque-Bernalillo County, New Mexico

Some items and issues that surfaced in the state-of-the-art literature assessment were not applicable to the Southwest. For example, although outdoor spaces were provided and landscaped, they were not situated well in relationship to indoor spaces and were little used. A critical aspect in the success of outdoor spaces in senior centers appeared to be visual contact between indoor and outdoor areas; the clients want to see and be seen, rather than be isolated from the main activities of the center. Another feature that diverges from findings in the literature was the emerging trend of the provision at senior centers for the physical fitness of senior citizens.

Concerns for future projects included the fact that most client organizations have limited experience with the process of POE and programming. As a result, these organizations may need evidence of the usefulness of POE efforts and of their value in improving future designs. While the reported project had a relatively small budget, part of its success may have been because the local Office of Senior Affairs supported the effort. In addition, many senior center clients had ample time to devote to project participation during the POEs and meetings in preparation for the generic senior center program.

ACKNOWLEDGMENTS

Thanks are owed to the members of the Senior Center Ad Hoc Planning Committee, as well as Ginger Grossetete, operations manager of the Albuquerque-Bernalillo County Office of Senior Affairs, who provided expertise and time for this project. The authors also thank Architectural Research Consultants, Inc., Richard R. Pugh (now Richard R. Pugh and Associates, AIA), and Barker/Bol & Associates of Albuquerque, New Mexico, for the collaborative spirit that helped make this project a success.

9. Investigative POE of the Agricultural Sciences Building - South, University of Kentucky

PROJECT SYNOPSIS

The purpose of this investigative POE was to assess building performance for the owners and occupants ten years after the building was completed and considered to be state-of-the-art. The purpose was also to explore in depth and with a high degree of accuracy a means of achieving better building utilization, ways to improve the building-delivery process for similar buildings in the future, and hypotheses about the effect of the physical design on building occupants. Another objective was to familiarize the project architect

1. Project Name:
   Investigative POE of the Agricultural Sciences Building—South, University of Kentucky

2. POE Level of Effort:
   Investigative POE

3. Purpose/Use:
   Verification of design premises, for example, flexibility in laboratories; POE training of staff

4. Building Type:
   Agricultural science laboratory and office building, University of Kentucky

5. Building Size:
   111,800 sq. ft.

Table 9-1. Project overview.

(continued)

6. Building Age:
   10 years (at time of evaluation)

7. Building Location(s):
   Lexington, Kentucky

8. Client:
   Bickel Gibson Architects, Louisville, Kentucky

9. Important Factors Analyzed:
   Technical
   HVAC, electrical, and structural soundness
   Functional
   Adequacy of space and environment; flexibility; access for the disabled
   Behavioral
   Security; orientation; sociometrics; psychological comfort

10. Methods Used:
    Data collection
    Interviews, questionnaire survey, observation, archival research, photography
    Data analysis
    Frequencies, means, chi-square test; canonical, regression, and factor analyses

11. Project Personnel:
    1 architectural researcher; 1 architectural principal; 1 junior architect; staff support; statistical consultant

12. Project Duration/Man-hours:
    12 weeks/320 hours

13. Project Cost:
    $15,000

14. Major Lessons Learned:
    Politics can severely alter and generate inappropriate design concepts; design of laboratory flexibility withstood test of time and change

15. Benefits to Client:
    Familiarization and training of staff in POE methodology; preparedness and success in obtaining major new commission from client organization

Table 9-1. (Continued)

with the POE process and to have the architect's staff gain some proficiency in the use of POE through involvement in the evaluation.

The results of this evaluation point to important factors in three areas that influence building performance and quality: the effect of early design and policy decisions and the building delivery process on building performance; the relationship between organizational change and building performance over time; and the effectiveness of specific performance criteria in meeting the needs of different occupant groups. A project overview is given in table 9-1.

INTRODUCTION

This investigative POE focused on the current use and objectives of the clients, who were the building's architect and the organization occupying the building at the time this POE was conducted. Consequently, information on evaluation criteria was gathered for conditions prevailing ten years after the building was first commissioned. While the initial design intentions were examined, they were primarily of historical interest.

The scope of the POE was limited by budgetary and time constraints. Because the building being evaluated was a highly specialized research laboratory, most of the findings and recommendations apply to this building type only. State-of-the-art information and evaluation criteria for the building type were difficult to obtain and study within the time frame of the project. With a twelve-week deadline, there was also limited time for the pretesting of data-gathering instruments.

In addition, the POE was conducted during the summer months, limiting the participation of student occupants of the building in data collection. However, since the building is primarily a research facility and the number of student users relatively small, issues related to seasonal changes were not considered critical.

BUILDING DESCRIPTION

The Agricultural Sciences Building - South centralized previously scattered animal science research activities at the University of Kentucky. The building consists of two major parts—a podium level and a tower. The podium level (37,000 square feet) includes facilities for animal care, dairy-products processing, meat processing, and a lecture and demonstration auditorium. The tower (74,000 square feet on ten levels) includes three classrooms, three teaching laboratories, twenty-seven research laboratories, 102 offices, and a mechanical / storage penthouse.

Among primary concerns of the client organization were isolation of the odor-producing areas and maximum flexibility of the laboratory spaces to accommodate frequently changing research programs. Campus-planning considerations and a limited site resulted in the tower solution—making the building the terminus and focal point for a proposed (but never implemented) elevated walkway leading from the central university campus to the north.

Flexibility, one of the major requirements of the building, meant providing for rapid modification and economical rearrangement of laboratory equipment and services. This was achieved by locating services in an interstitial space and intermediate service floor above the laboratories so that workers could easily make changes while minimizing interference with researchers' activities.

Figure 9-1. Main elevation.

By accommodating air-conditioning ducts and piping in concrete window spandrel beams and excluding above-ceiling mechanical work, a nine-foot floor-to-floor height was achieved in the office areas. This enabled two levels of offices to be provided for each level of laboratories.

The design for the main elevation of the building (fig. 9-1) won the 1975 Concrete Reinforced Structures Institute (CRSI) award. For a site plan and typical floor plans of the Agricultural Sciences Building - South, see figures 9-2 through 9-4.

1. PLANNING THE POE

The three-phase process as outlined in the POE process model in chapter 4 was used in this case study.

1.1. Reconnaissance and Feasibility

The project was initiated by Bickel-Gibson Architects of Louisville, Kentucky, the architectural firm that had designed the Agricultural Sciences Building - South. Architectural Research Consultants was retained by the project architect to conduct a post-occupancy evaluation between June and August of 1982.

Archival research and interviews with selected individuals helped reconstruct the program of the building during its initial planning stages. This included a review of the major decisions that had then been made, the changing occupancy of the building over the years, safety records and accident reports, and major systems concepts such as the air-handling and structural systems. The architects' project file was reviewed for minutes of meetings with representatives of the client organization. Sketches of schematic design concepts (fig. 9-5) and other documentation that might reveal why and how early design decisions were made were also reviewed.

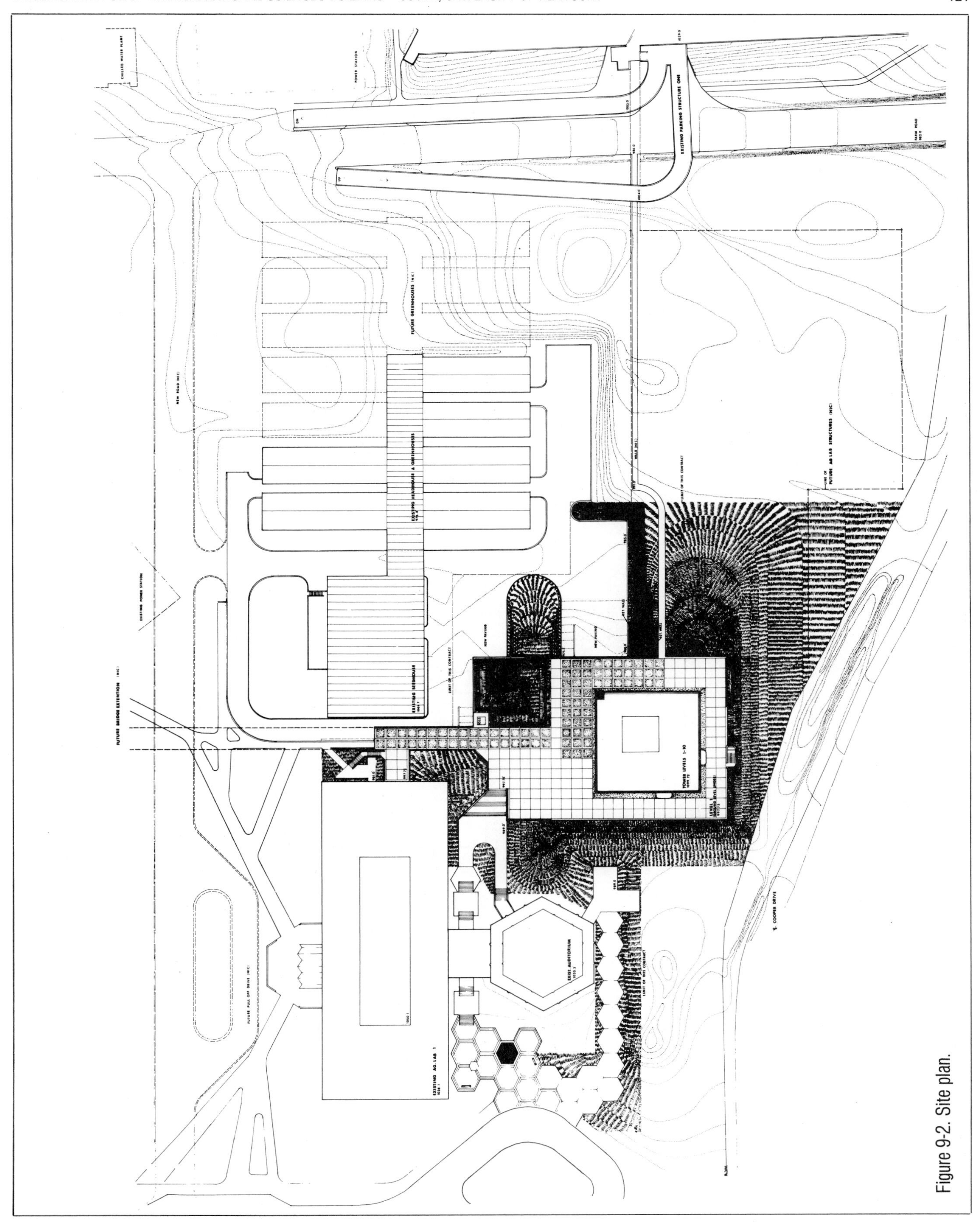

Figure 9-2. Site plan.

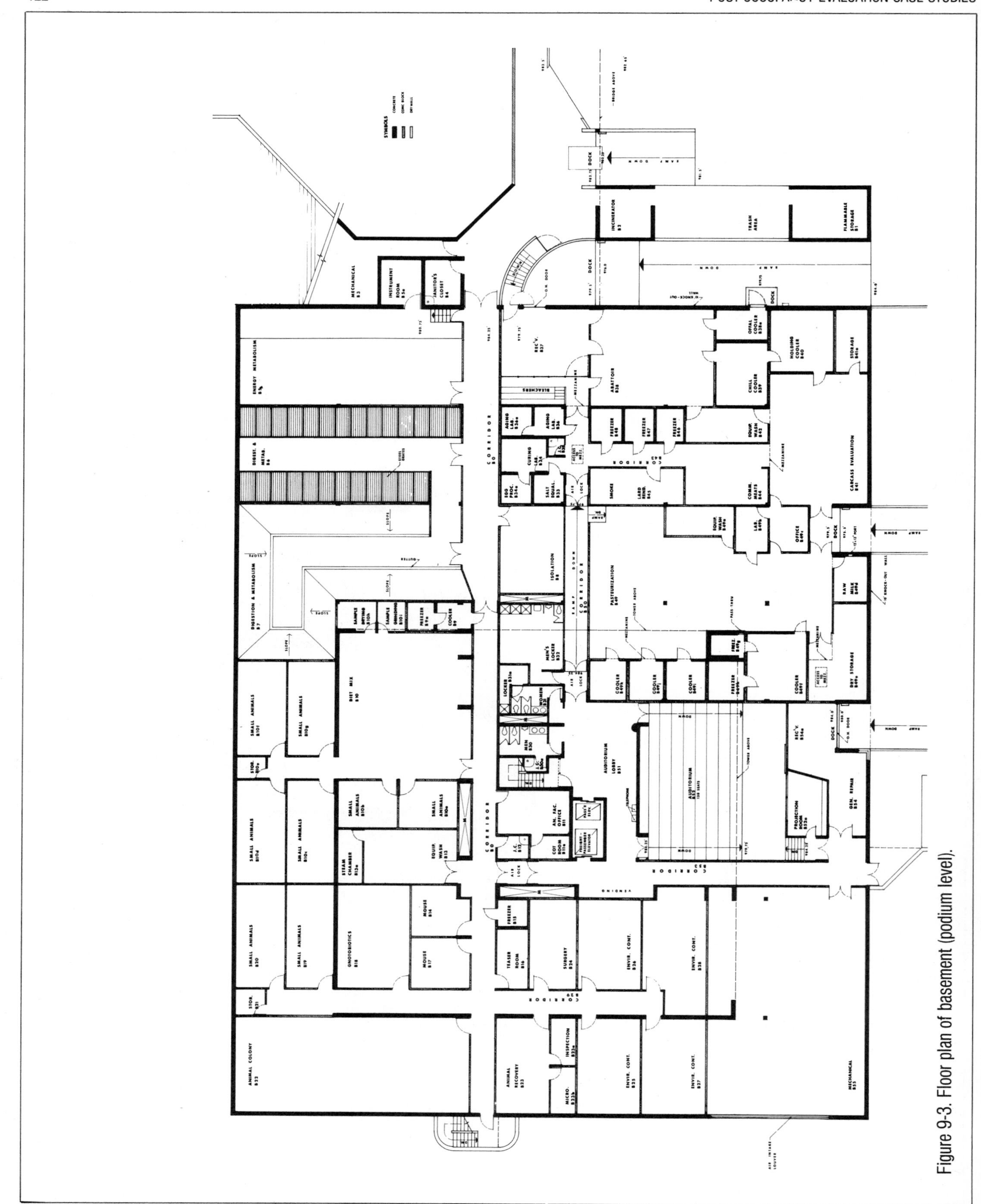

Figure 9-3. Floor plan of basement (podium level).

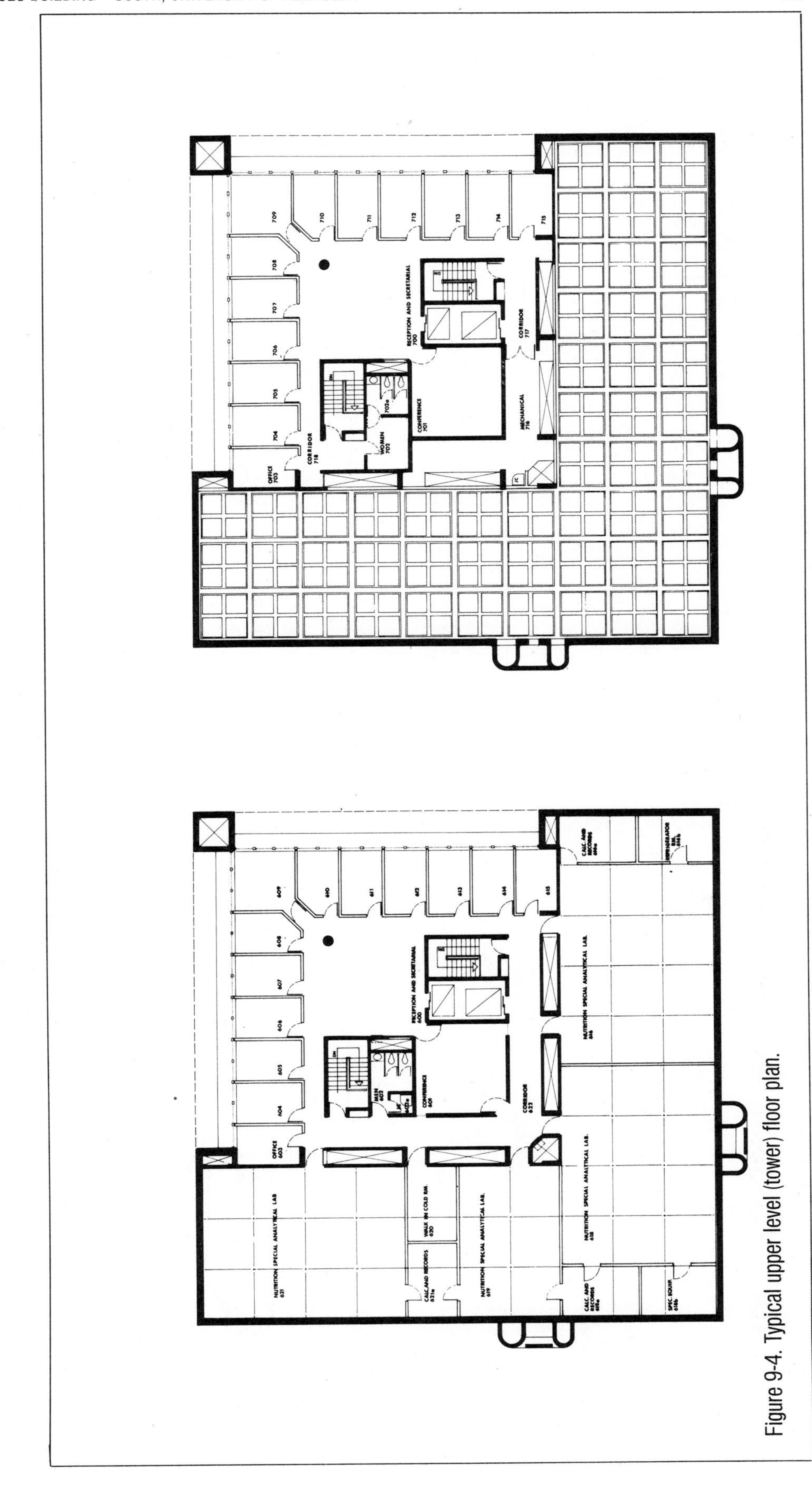

Figure 9-4. Typical upper level (tower) floor plan.

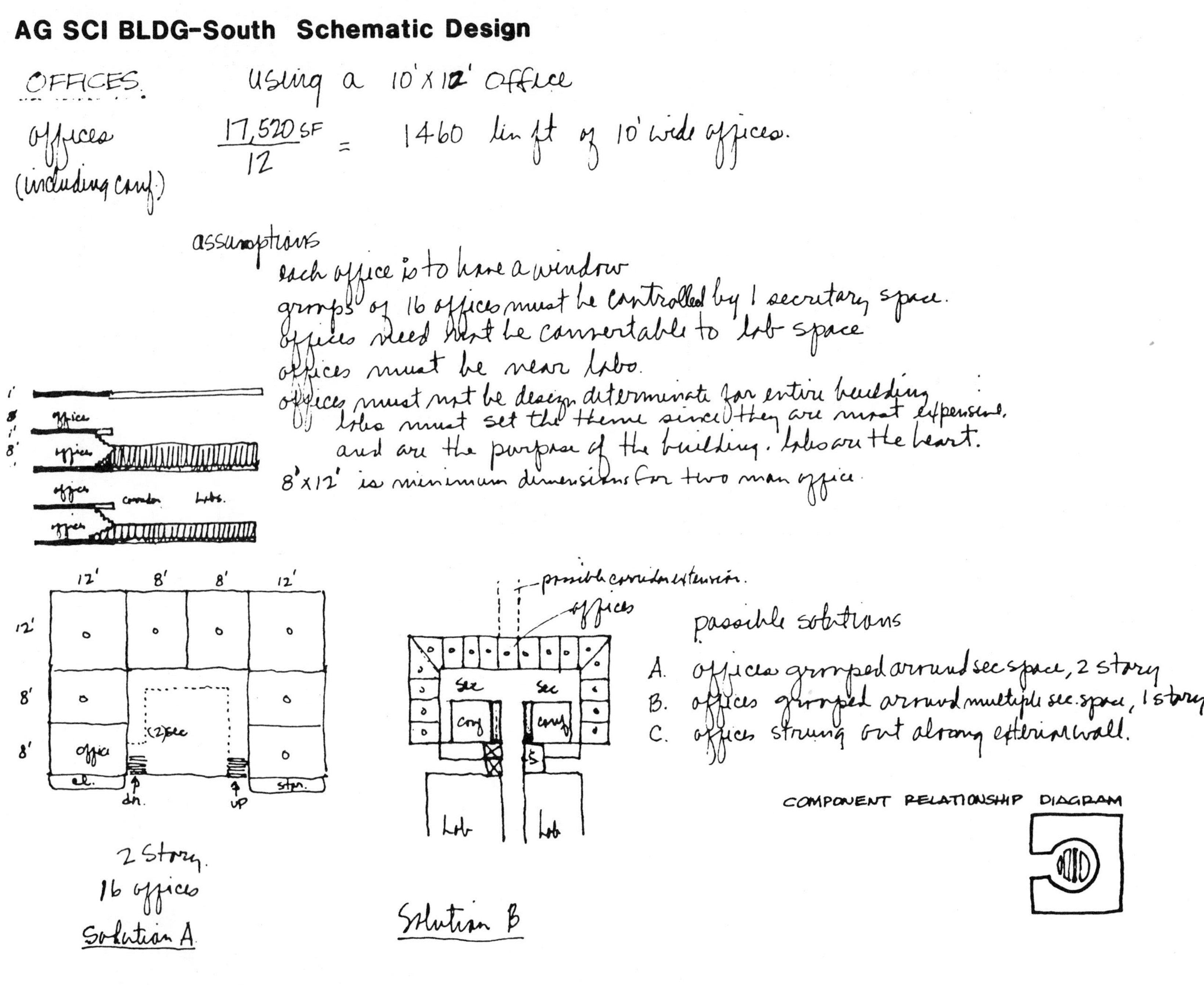

Figure 9-5. Early design concept sketches.

1.2. Resource Planning

The client organization, the College of Agricultural Sciences at the University of Kentucky, was visited by a team representing the project architect and the POE consultants, and the objectives and procedures of the POE were explained to a group of administrators and faculty. The cooperation and support of staff and other occupant groups at the facility being evaluated were obtained. Evaluators were given access to all spaces, permission to interview and survey building occupants, and permission to photograph and measure all aspects of the building and its use.

The POE team consisted of the project director (W. F. E. Preiser), a POE specialist from Architectural Research Consultants, the principal architect, and staff from the project architect's firm. The project director was responsible for overall coordination, project planning, training of support staff in data collection, analyzing data, and preparing the final report. Support staff in the project architect's office consisted of three people: a junior-level architect as liaison to the client organization, a coordinator for local data gathering, and a supervisor for additional support staff that included a secretary and a graphic artist. The project director managed the project from Albuquerque, New Mexico, requiring three site visits with approximately eighty man-hours on site. Interim communications were handled by regular and overnight mail and telephone. The project schedule is shown in figure 9-6.

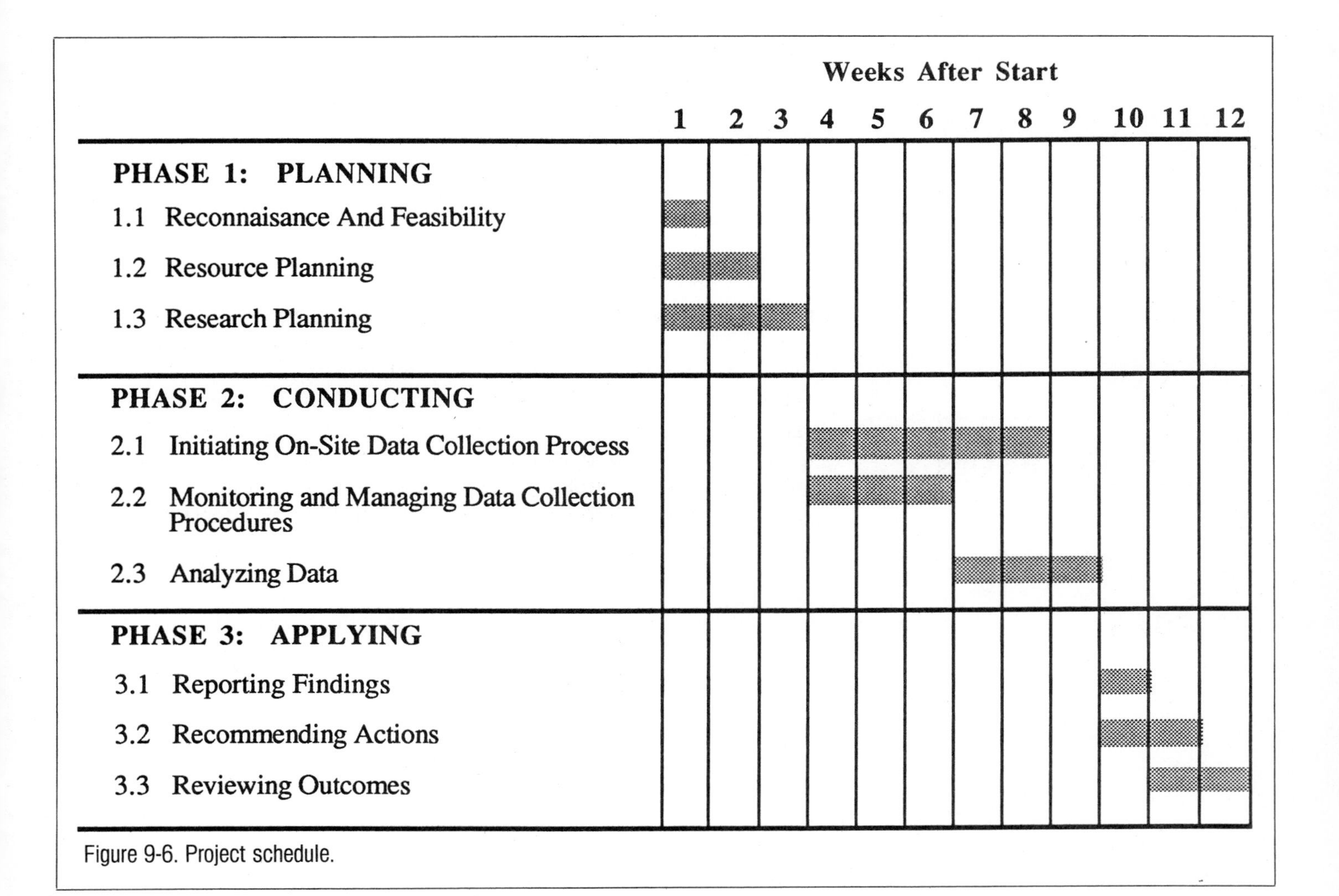

Figure 9-6. Project schedule.

1.3. Research Planning

Evaluation criteria were derived from four sources for the POE of the Agricultural Sciences Building - South:

- Expert judgment, developed for this specialized building type through the substantial experience of the evaluators in this field
- State-of-the-art literature and design guidelines for this building type (difficult to find in this case)
- State-of-the-art comparisons of the evaluated facility with similar recent ones around the country, using telephone interviews
- Experiences and subjective preferences of the building occupants, represented by surveys and observations

For some specialized spaces, no published criteria or literature could be found. In such cases, analogies with evaluation criteria for similar spaces were sought. Preliminary reconnaissance was made using a generic set of evaluation questions analogous to those in table 8-3. A multimethod approach was planned for data collection, including:

- Interviews and group discussions with representatives of the client organization
- Document and archival research
- Walk-throughs and inspections
- Still photography (slides and black-and-white prints)
- Direct observation
- A questionnaire survey of all occupant groups
- Detailed follow-up interviews

2. CONDUCTING THE POE

2.1. Initiating the On-site Data-collection Process

At this stage, data-collection instruments such as survey forms were prepared (fig. 9-7). The client organization was advised of the impending visit and survey. Interviewers were trained, and photography was scheduled. Detailed plans and schedules were made for observations, for interviews, and for any physical measurements that were to be made, such as sound measurements, building dimensions, and access for the disabled. Project participants included faculty, staff, and students of the College of Agricultural Science and the building's security personnel. Additional participants included the university architect and members of the university's central administration.

TO: OCCUPANTS AND USERS OF AGRICULTURAL SCIENCES BUILDING-SOUTH

DUE DATE: 25 June RETURN TO DEPARTMENT CHAIRMAN'S OFFICE BY NOON, FRIDAY.

RE: POST OCCUPANCY EVALUATION
AGRICULTURAL SCIENCES BUILDING, SOUTH
UNIVERSITY OF KENTUCKY

The Architect of the Ag Science Building, South and the Dean's Office wish to conduct a post-occupancy evaluation of the building. The purpose of this evaluation is to assess how well the building performs for those who occupy it in terms of the functions for which it was designed. The benefits of a post-occupancy evaluation include: consensus of good and bad features of the building for the betterment of future facilities; better building utilization; and improved building delivery process for future buildings.

Please respond only to those questions of the following survey that are applicable to you. Indicate your answers by marking the appropriate blanks with an "X."

1. In an average work week, how many hours do you spend in the following types of spaces:

Hours	Ground-level labs	Classrooms/ Auditorium B-52	Offices	Upper-level labs	Others, specify
0–5	( )	( )	( )	( )	( )
6–10	( )	( )	( )	( )	( )
11–15	( )	( )	( )	( )	( )
16–20	( )	( )	( )	( )	( )
21–25	( )	( )	( )	( )	( )
26–30	( )	( )	( )	( )	( )
31–35	( )	( )	( )	( )	( )
more than 40	( )	( )	( )	( )	( )

Key for the following rating questions: EX = excellent quality
G = good quality
F = fair quality
P = poor quality

2. Please rate the overall quality of the following areas relating to the building.

	EX	G	F	P
a) Ground-level laboratories	( )	( )	( )	( )
b) Upper-level laboratories	( )	( )	( )	( )
c) Offices	( )	( )	( )	( )
d) Classrooms/Auditorium (B-52)	( )	( )	( )	( )
e) Secretarial areas	( )	( )	( )	( )
f) Restrooms	( )	( )	( )	( )
g) Storage	( )	( )	( )	( )
h) Elevators	( )	( )	( )	( )
i) Stairs/corridors	( )	( )	( )	( )
j) Parking	( )	( )	( )	( )
k) Other, specify;______	( )	( )	( )	( )
______	( )	( )	( )	( )

Figure 9-7. Questionnaire survey form.

(continued)

3. Please rate the quality of the Ground-Level Laboratories in the terms of the following:

	EX	G	F	P
a) Adequacy of space	( )	( )	( )	( )
b) Lighting	( )	( )	( )	( )
c) Acoustics	( )	( )	( )	( )
d) Temperature	( )	( )	( )	( )
e) Odor	( )	( )	( )	( )
f) Esthetic Appeal	( )	( )	( )	( )
g) Security	( )	( )	( )	( )
h) Flexibility of use	( )	( )	( )	( )
i) Other, specify;______________	( )	( )	( )	( )
______________________	( )	( )	( )	( )

4. Please rate the quality of the Upper-Level Laboratories in the terms of the following:

	EX	G	F	P
a) Adequacy of space	( )	( )	( )	( )
b) Lighting	( )	( )	( )	( )
c) Acoustics	( )	( )	( )	( )
d) Temperature	( )	( )	( )	( )
e) Odor	( )	( )	( )	( )
f) Esthetic Appeal	( )	( )	( )	( )
g) Security	( )	( )	( )	( )
h) Flexibility of use	( )	( )	( )	( )
i) Other, specify;______________	( )	( )	( )	( )
______________________	( )	( )	( )	( )

5. Please rate the overall quality of the office areas in terms of the following:

	EX	G	F	P
a) Adequacy of space	( )	( )	( )	( )
b) Lighting	( )	( )	( )	( )
c) Acoustics	( )	( )	( )	( )
d) Temperature	( )	( )	( )	( )
e) Odor	( )	( )	( )	( )
f) Esthetic Appeal	( )	( )	( )	( )
g) Security	( )	( )	( )	( )
h) Flexibility of use	( )	( )	( )	( )
i) View to the outside	( )	( )	( )	( )
j) Visual privacy	( )	( )	( )	( )
k) Ceiling height	( )	( )	( )	( )
l) Storage space	( )	( )	( )	( )
m) Others, specify;______________	( )	( )	( )	( )
______________________	( )	( )	( )	( )

6. Please rate the qualities of materials used in this building *as a whole.*

	EX	G	F	P
a) Floors	( )	( )	( )	( )
b) Walls	( )	( )	( )	( )
c) Ceilings	( )	( )	( )	( )

Figure 9-7.

7. Please rate the quality of the building *as a whole* in terms of the following:

	EX	G	F	P
a) Esthetic quality of the exterior	( )	( )	( )	( )
b) Esthetic quality of the interior	( )	( )	( )	( )
c) Amount of space	( )	( )	( )	( )
d) Environmental qualities (lighting, acoustics, temperature, etc.)	( )	( )	( )	( )
e) Proximity to colleagues	( )	( )	( )	( )
f) Adaptability to changing uses	( )	( )	( )	( )
g) Security	( )	( )	( )	( )
h) Maintenance	( )	( )	( )	( )
i) Relationship of spaces/layout	( )	( )	( )	( )
j) Quality of immediate work environment	( )	( )	( )	( )
k) Other, specify;__________	( )	( )	( )	( )
__________	( )	( )	( )	( )

8. Please select and rank in order of importance five (5) qualities from question #7 that contribute to a work environment that has the quality you prefer.

1.__________
2.__________
3.__________
4.__________
5.__________

9. Please identify important facilities which are currently lacking in the Ag Science Building, South.

10. Please make any other suggestion you wish for physical or managerial improvements in the building.

11. Comparing Ag Science South (II) and Ag Science North (I), which working environment would you prefer?

Ag Science South (II) ______
Ag Science North (I) ______
Please explain why.

12. Please circle the floor where your primary work area is located.

G 1 2 3 4 5 6 7 8 9

Figure 9-7. *(continued)*

13. What is your job title or position? ____________

14. Are you part-time ________ or full-time ________?

15. Sex: M________ F________

16. Age: Under 20 ________ 41–50 ________
20–30 ________ 51–60 ________
31–40 ________ Over 60 ________

Please return these questionnaires to the Department Chairman's Office by *FRIDAY, 25 June 1982 at NOON.*

Thank you for your time and effort.

Figure 9-7. (Continued)

Document and archival data collection involved information related to both client and building type, including:

- Client statement of purpose
- Organizational chart and staffing
- Initial program for the building
- As-built floor plans
- Space assignments and use schedules
- Building-related accident reports
- Maintenance and repair records
- Energy audits or review comments from the university heating / cooling plant manager
- Self-study and other evaluation materials
- Security records
- Cost of modifications

Identification and review of recent, similar facilities, through telephone interviews with respective user agencies, were used to help develop criteria for this building type. Data included: a review of programs, plans, and other pertinent state-of-the-art documentation; evaluation of design concepts including office-to-lab proximity, horizontal versus vertical space organization; location of animal areas; secretarial area layout; and fenestration.

2.2. Monitoring and Managing Data-collection Procedures

Several measures were taken to assure that valid and reliable data were collected on site. In terms of direct observation and still photography, inter-observer reliability checks were carried out to ensure that identical observations were in fact recorded by different members of the field research team.

Interviews were tape-recorded and strictly followed a predetermined schedule of questions presented in identical order to each of the interviewees. Issues were covered in a systematic manner while permitting room for unsolicited or open-ended responses from the interviewees. Likewise, survey questionnaires were administered in such a manner that the predetermined sampling strategy was followed exactly; a representative response from the stratified sample of respondent groups such as students and staff was facilitated. Thanks to effective communication with all respondent groups, the response rate was excellent.

Communication with administrators, department heads, and other key persons in the client organization were frequent, both over the telephone and through face-to-face on-site meetings. Thereby it was possible to focus the client organization and the respondents on the purpose of the POE and specifically the kind of care that was required in employing the various data-gathering methods.

2.3. Analyzing Data

INTERVIEWS

Interview data were categorized and analyzed for frequency, content, and importance of issues mentioned by respondents (see fig. 9-8). These data were used in structuring questionnaire items for the survey that followed.

OCCUPANT SURVEYS

Data were analyzed for regular building occupants (and informally for visitors). The sample consisted of 101 respondents in the Department of Animal Science (out of a possible 160) plus 33 respondents from the Department of Agricultural Economics (out of a possible 75).

Frequencies, means, and standard deviations were used for analyzing the rating data from the survey. Stepwise regression (a statistical technique that pinpoints closely related variables when many variables are used) was employed to identify relationships among group-membership variables (position, age, sex, primary work area, and so on) and judgments of various aspects of building performance.

More sophisticated analytical techniques were also planned for and used with the help of a statistical consultant in this area. Factor analysis was used to analyze ratings of specific questionnaire items. This technique reduced the original set of ten variables (quality ratings for areas of the building) to three variables to identify underlying patterns or dimensions that explain respondents' perceptions.

Another technique, canonical correlation analysis, was used on other questionnaire items to examine how factors of group membership such as position and primary work area influenced respondents' evaluations of the quality of selected areas in the building and the building as a whole. Certain preliminary hypotheses were generated. Finally, chi-square tests were conducted to identify possible differences in preference as they related to factors of group membership (position, department, and primary work area).

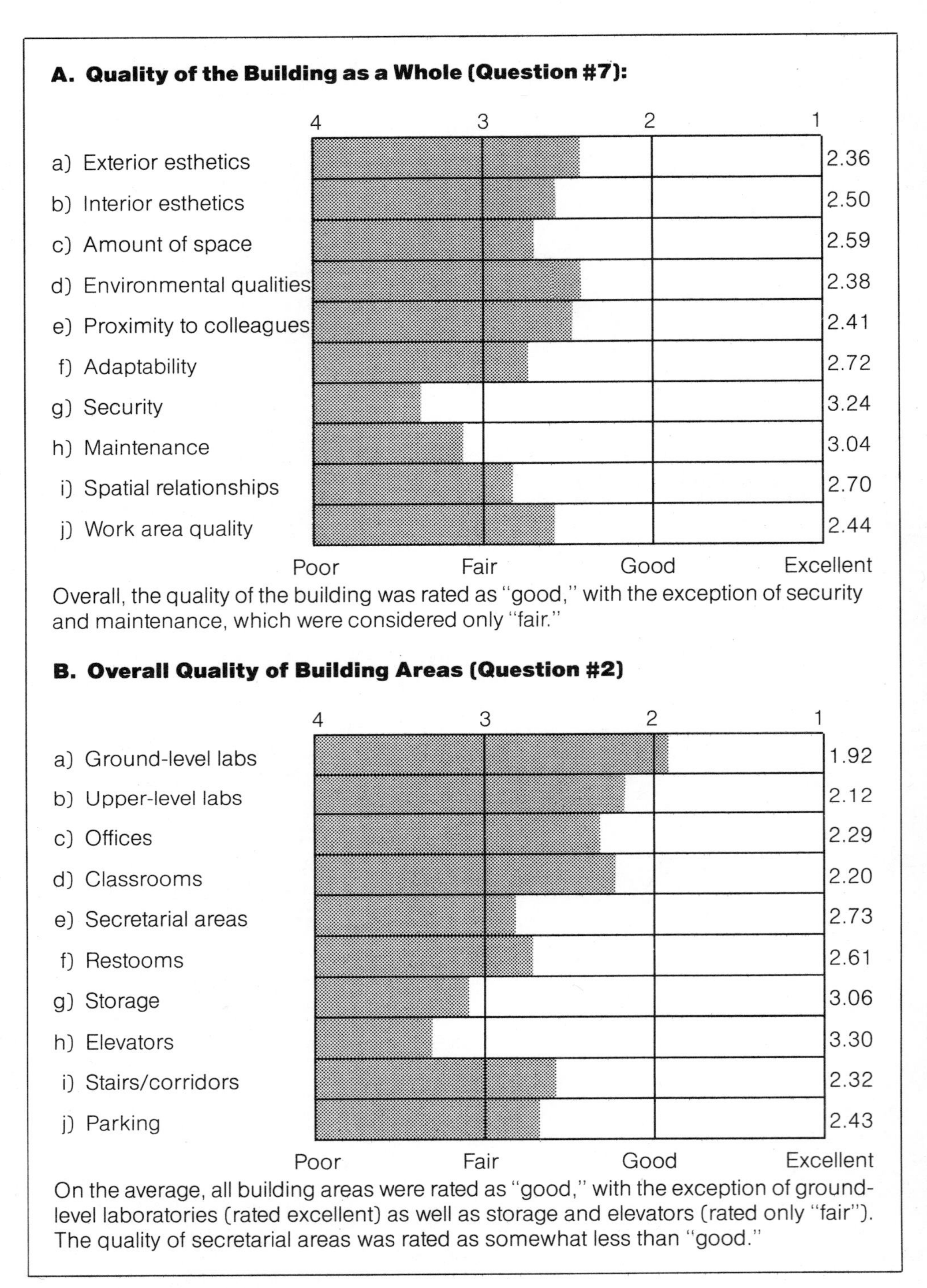

Figure 9-8. Example of data summaries.

OBSERVATION AND PHYSICAL INSPECTION

The data obtained were tabulated and categorized according to space type and the behaviors and physical phenomena observed. For example, variations in layout and use of secretarial areas on eight different floor levels were analyzed. Where appropriate, physical inspection data were documented in the form of annotated floor plans.

3. APPLYING THE POE

3.1. Reporting Findings

The multimethod data collection and analysis approach used in this POE proved to be an effective way to accurately assess the successes and failures of the building being evaluated in a relatively short time. Through the initial, structured interviews, combined with the walk-throughs, the research team was able to identify major issues regarding the performance of the building, which pointed to possible shortcuts in evaluation methodology. Survey data confirmed these early findings and established certain causal relationships among physical environmental characteristics of the building, as well as occupant perceptions of its quality.

OVERALL BUILDING PERFORMANCE

After ten years of occupancy, the client organization considered the building functionally appropriate—especially in the laboratory areas, where extraordinary flexibility for adaptation to changing needs had been part of the building program and design concept. These findings were supported by occupant surveys, indicating "good" quality ratings for virtually all aspects of the building. An exception was ratings from the members of the agricultural economics department who were assigned to three floors of the building. They responded most critically to their office accommodations, which had originally been designed to meet the program requirements of personnel in the animal sciences department who required close proximity to laboratories. In some cases facilities that were originally provided for one use had been converted to other uses (for example, computer, conference, and storage rooms), or the number of occupants had increased beyond the original program intent.

DESIGN CONCEPTS

Some occupants criticized the decision to combine the office and laboratory tower with a large animal laboratory at the ground level because of occasional air-handling problems, mainly in keeping animal odor out of the tower. Despite conceptual studies and proposals by the architect to use a "horizontal" layout, as had been used in the older Agricultural Sciences Building - North, site- and campus-planning considerations prevailed, even though comparison with state-of-the-art facilities elsewhere had indicated that the horizontal layout was preferred in agricultural science laboratories.

The proximity of faculty offices to lab areas, common in most current research facilities, was considered very convenient and functional, although some occupants felt that the vertical separation of departments reduced informal contacts among faculty and staff. These feelings were especially prevalent among faculty who had experienced the older, horizontal building.

The ninth-floor location of the administrative office of the agricultural science department was not considered ideal because it contributed to unnec-

Figure 9-9. Secretarial areas on the third and fourth floors.

Figure 9-10. Open guardrail in main lobby area.

Figure 9-11. Damage due to chemical fumes and liquids.

Figure 9-12. Interstitial service space in laboratories.

Figure 9-13. Crack in ceiling showing minor water damage.

essary vertical traffic through the building and congestion in the elevators. A first-floor location would have been more appropriate.

The openness of secretarial areas to the elevator (fig. 9-9) contributed to perceived security problems, although actual security was better than elsewhere on the university campus, as evidenced by a check with campus security regarding the occurrence of theft.

The location of restrooms for men and women on alternate floors (due to space limitations) was criticized as being unusual and inconvenient.

As was expected, the range of findings was considerable—extending from technical, health, safety, and security issues to functional performance and a number of psychological, comfort, and satisfaction concerns. For example, there were safety problems with open guardrails in the main lobby area (fig. 9-10) and unsafe storage of liquids and gases in the laboratory area, where chemical fumes and liquids had damaged the floor, walls, and ceiling (fig. 9-11).

Functional performance was documented, for example, in a photograph of interstitial service space in laboratories that permitted easy adaptation to changing technology over the years (fig. 9-12). Finally, technical performance evaluation was also documented; for example, in figure 9-13 minor water damage is seen in the ceiling, caused by a crack in the concrete beams.

3.2. Recommending Actions

POE SERVICE

One of the objectives of the project was for the architects involved to become proficient in using POE, and this was successfully accomplished. It was recommended that architects adopt POE as one of the services offered client organizations, and it was shown that, as the architect's staff gains experience, the POE process becomes quite efficient and less time-consuming. In addition, for architects who specialize in particular building types and are familiar with use-related issues in those buildings, POE will enable them to focus rapidly on critical evaluation items.

IN-HOUSE POE CAPABILITY

For a large institutional client organization such as a university, the creation of an in-house POE capability is recommended. Limited-scope POEs should be carried out continually as a staff function at three-to-five-year intervals. As a result, a changing organization may be able to ensure maximum utilization of space and limited resources.

In reviewing the outcomes of this POE, it is recommended that government agencies and large institutions create in-house staff capability to carry out POEs routinely and on an intermittent basis. For example, this could be accomplished through the university architect's office at the University of Kentucky, thus facilitating continuous feedback concerning the performance of the buildings in use. The emphasis of such POEs could vary depending on the information needs. For instance, space-utilization information is in constant demand by the space planners and could be obtained through POEs.

BUILDING USER MANUAL

Following the identification of a variety of use-related issues in the building, it was recommended that a user manual be written to cover all important aspects of building use, including fire and emergency procedures, seasonal adjustments, flexibility for change, allowable live loadings, and energy-conservation measures.

3.3. Reviewing Outcomes

METHODS

The methods employed were somewhat sophisticated for the singular POE case presented here. The methodology can be considered a pilot application of data gathering and analysis that could well be suited to cross-sectional, comparative evaluations of a representative sample of a given building type. The multimethod approach used in this POE could have been simplified and reduced in scope somewhat, especially as far as data-analysis techniques are concerned. Techniques employed included factor analysis, canonical correlation analysis, stepwise regression analysis, and chi-square tests of building preference (see glossary).

COST AND TIME EXPENDITURE

While the project extended over a period of twelve weeks, the actual expenditure of time and effort was considered reasonable, particularly since part of the project entailed the training of the project architect's staff in conducting POEs. The project architect considered the cost of $15,000 for this POE a good investment in possible future projects with the University of Kentucky. The POE effort was particularly worthwhile since it provided feedback to the architect on how well a building he had designed had performed for the client organization. It also gave him a considerable edge (and sound, research-based information) in the competition for the design commission for a new project, the Equine Research Center, a competition subsequently won by the project architect.

OVERALL FINDINGS

The Agricultural Sciences Building – South at the University of Kentucky rates highly in comparison with similar state-of-the-art facilities, particularly in terms of interior spaces and laboratory flexibility. However, the tower concept is partly responsible for the major design flaws that went undiscovered.

FEEDBACK

Feedback from the project architect regarding this POE highlighted one of the true benefits of POE:

> Since completing work on the post-occupancy evaluation of the University of Kentucky Agricultural Sciences Building – South, we have designed

another, major $7-million laboratory building for them which is dedicated to equine research.

The knowledge the POE gave us of design concepts that worked versus ones that were less successful reduced the need to reinvent the wheel. An added benefit was the trust that developed between Bickel-Gibson and the university, which ultimately led to greatly improved communications. They knew we cared.

James Gibson
Bickel-Gibson Architects
Louisville, Kentucky

ACKNOWLEDGMENTS

This POE was initiated and supported by Bickel-Gibson Architects of Louisville, Kentucky. Research and report results were guided by Dr. Wolfgang F. E. Preiser, director of research of Architectural Research Consultants in Albuquerque, New Mexico, with assistance in statistical analysis by Jerry Manheimer from the University of New Mexico Department of Psychology. We appreciate greatly the support and work of David Heyne of Bickel-Gibson. Thanks are owed to the client, the University of Kentucky and especially the College of Agriculture, for its excellent cooperation.

10. Diagnostic POEs of Elementary Schools in Columbus, Indiana

PROJECT SYNOPSIS

The objectives of this diagnostic POE project were threefold: to provide results useful to the client school board, owner and operator of the facilities; to generate knowledge in the area of the elementary school design for use by design professionals and school facility managers; and to provide basic research into POE, including a rigorous investigation into all three categories of building performance—technical, functional, and behavioral.

The POE took about eighteen months to complete and included many types of data-collection methods: a literature search, instrumentation, surveys, inventories, behavioral observation, interviews, and the assessment of archival records. Hundreds of different measurements were taken, and many performance criteria were developed.

The results were reported in three volumes, now available as a single document (Rabinowitz 1975). Although there are a large number of specific results, some of the more provocative findings include the frequent use of nonclassroom areas (primarily corridors) for educational activities, the non-use of areas designed for special activities, flaws in roof and exterior walls, and the need for more and better-located storage areas.

1. Project Name:
   Diagnostic POEs of Elementary Schools in Columbus, Indiana

2. POE Level of Effort:
   Diagnostic

3. Purpose/Use:
   Evaluation for improvement of existing buildings; exploration of POE methodologies

Table 10-1. Project overview.

4. Building Type:
   Four elementary schools

5. Building Size:
   50,000 sq. ft. each

6. Building Age:
   16, 8, 4, 2, at time of evaluation

7. Building Location(s):
   Columbus, Indiana

8. Client:
   Bartholomew Consolidated School Corporation, Columbus, Indiana

9. Important Factors Analyzed:
   Technical
   Comprehensive
   Functional
   Comprehensive
   Behavioral
   Comprehensive

10. Methods Used:
    Data collection
    Literature review, instrumentation, observation, photography, inventory, survey
    Data analysis
    Comparison with existing standards; frequencies, means, percentage distribution

11. Project Personnel:
    1 university researcher; 1 research assistant; 4 consultants

12. Project Duration/Man-hours:
    16 months/18 months

13. Project Cost:
    $80,000

14. Major Lessons Learned:
    Frequent utilization of space for unintended purposes, inadequacy of existing spaces

15. Benefits to Client:
    Improved space standards, for example, for storage; unexpected occurrence and discovery of minor technical failures; surprising traditional behavior in open plan schools; awareness of effect of school environment on student behavior; repair of minor problems; applicability of findings to future schools

Table 10-1. (Continued)

1. PLANNING THE POE

1.1. Reconnaissance and Feasibility

The project was initiated by the principal investigator, a faculty member of the School of Architecture and Urban Planning at the University of Wisconsin–Milwaukee, and was supported by the School Board of Bartholomew County, Indiana. Located in Columbus, Indiana (population 25,000), the county seat, are the headquarters of the Cummins Diesel Company, the world's largest manufacturer of diesel engines. Many Cummins executives and skilled workers live in that city. Through the generosity of a number of foundations in Columbus that advocate excellence in architecture, Columbus also has as many buildings designed by nationally prominent architects as do most major cities. The school board, interested in this pioneering POE, also obtained support from one of these local foundations.

While Columbus had many schools designed by prominent architects, elementary schools were chosen for this POE for five primary reasons:

1. There were many elementary schools from which to choose, while there were very few middle schools and high schools because they tend to have large student bodies.
2. The scale of the elementary schools was relatively modest. Each had an area of about 50,000 square feet, and this size made a comprehensive POE feasible.
3. The elementary schools were neighborhood-based, and many had similar student populations.
4. The greatest variety of design solutions occurred at the elementary school level, and they represented a variety of educational philosophies as well.
5. In terms of study options, the elementary schools tended to have much simpler designs than upper-level schools, despite differing philosophies. Since much of the research presented in this POE project was unprecedented at the time, there was a need to keep the project methodologically straightforward.

Basic demographic data from the elementary schools and their neighborhoods were examined. Data such as parents' occupations, income, and race were examined, as were size and architectural attributes of the schools, student population, rate of absenteeism, and teacher experience. The four schools chosen were quite similar: they were of the same size (about five hundred students) and had students from similar socio-economic backgrounds. The buildings chosen had strong architectural and morphological attributes that promised interesting implications in the evaluation.

The principal investigator also chose schools in which administration and staff seemed eager to cooperate in the evaluation. This was particularly important in a diagnostic POE because the numerous observations, meetings, interviews, and surveys would require a good deal of time from the building administrators and occupants.

1.2. Resource Planning

Resource planning was a critical part of this ambitious POE, and a number of steps were taken that in retrospect were important not only in resource planning, but in terms of enhancing the quality of the project.

1. A lean project staff aided not only the allocation of the scarce project funding, but also the project's quality. The principal investigator and a graduate assistant/project manager ran the project with several consultants in specific areas of expertise and temporary help when needed.
2. A project manager with excellent writing, analytical, and graphic skills was a key factor in the POE's success. This assistant had important responsibilities throughout the project.
3. Most consultants in the fields of acoustics, lighting, education (psychology and curriculum), and construction were university faculty members.
4. Local on-site observers proved to be quite effective for a number of reasons: they were all parents of local schoolchildren and thus were motivated to participate in the POE and they had local knowledge and experience.
5. Fewer but more productive trips were made to the POE sites. Typically, intensive site visits lasted two to four days, and work often continued far into the night.

Estimates of personnel time and expenses were made for the entire project and were reviewed every two months. The estimates were generally accurate for the data-collection phase but were low for the data-analysis and report-writing phases.

Testing equipment had to be acquired for the technical testing; some of this was borrowed from university departments. For example, the physics department had sound-level meters and photometers, and the physical plant department had recording thermometers and a velometer, to measure air velocity. An effective device was obtained from the university's risk-management department to test the slip resistance of wet and dry floors.

In a preliminary way, resource planning also concerns the type of evaluation methods that will be used, particularly in data collection and analysis concerning behavioral elements of performance. Although direct observation is the most resource-intensive data-collection method, when compared to questionnaires or interviews, its reliability was the most important consideration in this case.

As discussed in chapter 6, tape recorders and cameras are invaluable tools throughout a POE. For this entire project some fifteen hundred slides documenting the project were taken.

1.3. Research Planning

During the reconnaissance and research-planning steps, a literature search was conducted on performance criteria. In the process of locating literature material many information gaps were identified. For the technical elements of performance, many laboratory tests were adapted for on-site conditions, and a technical field test manual was produced. Satisfactory performance criteria were found for a number of functional elements while behavioral elements of performance yielded little in terms of relevant research or criteria. This last area of concern was the most exploratory one in a POE that investigated the relationships between the environment and the behavior of the occupants.

Many technical criteria and tests were identified in the literature of national organizations, such as the National Bureau of Standards or the American Society for Testing of Materials (ASTM). Sometimes criteria originated from very specialized sources. For instance, criteria for blackboard glare came from a manufacturers' association. In some cases, criteria were developed in the course of the project, based on primary research when existing criteria were inadequate. For example, for classroom storage the best criteria found in the literature were exceeded by the storage requirements of all four schools evaluated.

Forms for data collection and aggregation were developed beforehand and pretested both in Milwaukee and on site. Data were reviewed by the researchers as they were collected in the field and before they were returned to the research office. This procedure helped ensure that, if errors did occur, they would not be repeated.

Although most elements of performance were evaluated, some areas were particularly emphasized as a result of early discussions with the client and responses to questionnaires administered to teachers in the four schools.

Detailed research schedules were developed for the project's assistants and consultants. Behavioral aspects were given special attention. Faculty consultants from the University of Wisconsin School of Education helped both in scheduling and setting up the analysis for this component of the POE. Behavioral observations spanned four months and took place according to a representative time-sampled schedule that covered every time of the week in the evaluated school buildings.

Three trips for data collection involved at least five persons and a number of instruments. A central but out-of-the-way location in a corner of the teachers' lounge was used as a temporary field office.

2. CONDUCTING THE POE

2.1. Initiating the On-site Data-collection Process

Most of the technical and functional data were collected in the evenings and on weekends. The scale of the project, in terms of both the number of performance elements to be evaluated and the areas sampled, would have interfered with institutional activities during school hours.

2.2. Monitoring and Managing Data-collection Procedures

The practice runs for data collection on the behavioral elements of performance were useful in ameliorating the reactions of the occupants to data collection. The observers not only made notations but also photographed every space sampled. Within two days the students learned to ignore the observers, who had been chosen and trained to be as unobtrusive as possible.

2.3. Analyzing Data

Data analysis was planned in earlier phases of the POE, but surprises still occurred. The frequent use of the corridors as educational settings was unexpected and required a modification of data collection and analysis procedures in order to focus more closely on those spaces. Regarding technical performance, poorly designed roofs did not leak while well-detailed ones did. In many ways, the open plan schools were more conservative than the more traditional ones. Were the data correct? More analysis was needed.

Often, the findings from POEs have been so strong that many times "coarse" measures are quite appropriate and are suggested by the faculty consultants from the School of Education. A straightforward attitude toward analysis was adopted, primarily using descriptive statistics.

Most technical and functional evaluations used direct comparisons between actual building performance measures and performance criteria, for instance, in evaluating acoustical performance. In evaluating ambient sound levels, the measures in the classrooms tested in each school were compared to a well-known standard. Regarding functional elements of performance, a test was conducted on room darkening (for the viewing of slides or films) and was compared to recommended levels.

Concerning behavioral performance elements, there were few criteria in existence. The goal of this POE was to point out relationships between design attributes and the behavior of building occupants. For instance, a study of student posture based on observational data strongly linked the provision of carpeting in the schools with students sitting on the floor. In one school, carpeted steps designed for sitting were an important innovation that worked well—more than 30 percent of the students sat on the "floor" of this school, based on more than two thousand observations of student postures in this school alone.

TECHNICAL EVALUATION: REVERBERATION TESTING

Reverberation is an acoustical property of a room—specifically, it is the length of time that a sound persists in the space as a result of its being reflected off of the room's surfaces.

Reconnaissance of the schools involved obtaining as-built plans and specifications and visiting the four buildings. Listening to the spaces in the buildings as well as examining their volume and materials provided a preliminary list of spaces to be tested for reverberation. A field test method was developed using the standard 60-decibel decay time as the test criterion (fig. 10-1). Decay time is the time it takes for a sound to diminish a specific amount. A measured decay of 60 decibels is called the *reverberation time.* Because of

- Ambient sound level, ± dbA, in source room with source on;
- Ambient sound level, ± dbA, in receiving room with source on.

TEST # 3: Control Reverberation Within Spaces

Test Method: Determine past performance if possible. Test by using 18-inch-diameter balloons to provide an instantaneous and loud sound source. A tape recorder specifically modified for the purpose records the reverberation test. Two trials are recorded in each space. The tape is analyzed in a laboratory to determine the reverberation time. Measurement is made at least 30 inches from any reflecting surface.

Since the equipment used in this test is quite expensive and sophisticated and since reverberation detrimental to normal speech should be heard using the balloons, it is possible to burst balloons and simply note the discernible reverberation if any.

Measures: Reverberation times

- Reverberation times for the following frequencies: 125, 250, 500, 1,000, and 2,000 HZ.

TEST # 4: Control Mechanical Systems' Noise

Test Method: Determine past performance if possible. Some as test #1. Readings are taken with lighting and mechanical systems turned off and turned on.

Figure 10-1. Reverberation testing.

the sophistication of the reverberation test, an acoustical expert was consulted. A literature search was conducted for acoustic performance criteria.

The evaluators conducted preliminary tests on selected small and large spaces, using sensitive sound-level meters, large balloons, and stopwatches. The acoustical consultants—Bolt, Beranek, and Newman—conducted, recorded, and analyzed the final tests. The results were similar in the large spaces that had significant reverberation times easily detected by preliminary estimates. In classroom spaces with reverberation times of about one second, it was difficult to conduct accurate field testing without sophisticated equipment for data collection and analysis.

FUNCTIONAL EVALUATION: STORAGE

Storage spaces were photographed during the reconnaissance step. Working drawings that documented storage spaces were obtained, and it became apparent that significant amounts of "unofficial" storage existed.

Few criteria for storage in schools were found in the literature, and those that existed were outdated and inadequate for any of the schools evaluated. Therefore, criteria were developed by the investigators based on measures of actual classroom storage.

Budgeting considerations required that only second-, fourth-, and sixth-grade classrooms be evaluated. The measurement of storage capacity and use was planned for after-school and evening hours. The actual use of stored materials by teachers and students was not observed; however, the teachers were surveyed about how frequently they used different classroom materials.

In conjunction with field measures and worksheets, photography was systematically used to document storage in the classrooms sampled. The performance of immediate-, medium-, and long-term storage accessibility needed by teachers and students was not part of the original evaluation process, and, therefore, remeasurement was required. Figures 10-2 and 10-3 illustrate the evaluation of storage in the schools.

STORAGE CAPACITY

Performance required: Provide adequate storage for classroom and school needs.

Method: A comparison of the 'official' storage provided and the actual use of storage - both official and unofficial. A detailed storage 'inventory' of most classrooms was made, including photo documentation.

Analysis:

STORAGE CAPACITY

	STANDARD*	P	R	S	M
STORAGE CAPACITY PROVIDED PER CLASSROOM (cu.ft.)	250 (approx)	146	155	182	350

*standard developed through analysis of existing conditions

The originally specified storage capacity has increased for each new school. This wide range of alternatives is a good 'experiment' against which to test the performance of storage capacity.

'Overflow' storage was consistently and obviously evident at the Parkside, Richards, and Smith Schools. Every cubic foot of storage space provided was brimfull as were numerous other locations. Shelves in many cases were literally deflecting with the weight of their loads. Most shelf storage was multilayered - the objects being piled atop one another also creating problems in organization and disarray.

At the Parkside and Richards Schools overflow storage took place on the floor, in cardboard boxes and on folding tables. Occasionally a storage cabinet would be brought in from home by a teacher. At Smith most overflow storage occurred in the nodes in metal cabinets and steel shelving though the 3'x 6' folding tables were again used in the classroom.

The Mt. Healthy School has sufficient built-in cabinetry and mobile cabinets, some of which went unused.

Findings: Based on these very consistent results we would recommend 250 cubic feet of storage area be provided for each classroom. Centralized storage needs would require an additional 50 cubic feet for each classroom.

SIZE OF STORAGE

Performance required: Provide storage for objects of various sizes.

Method: See 'Storage Capacity".

Analysis: There is a storage problem for all sizes of objects related to the lack of storage capacity. Universally displaced, however, are those large objects, say over 15 inches in all dimensions, for which the shelving systems and drawers make no provision. This includes:

most audiovisual equipment
screens and charts
instructional media kits and sets
globes
recess equipment
easels
large models (eg. clockface, earth and moon)
fish tanks
plants
fans
cardboard boxes of miscellaneous objects

Figure 10-2. Storage evaluation.

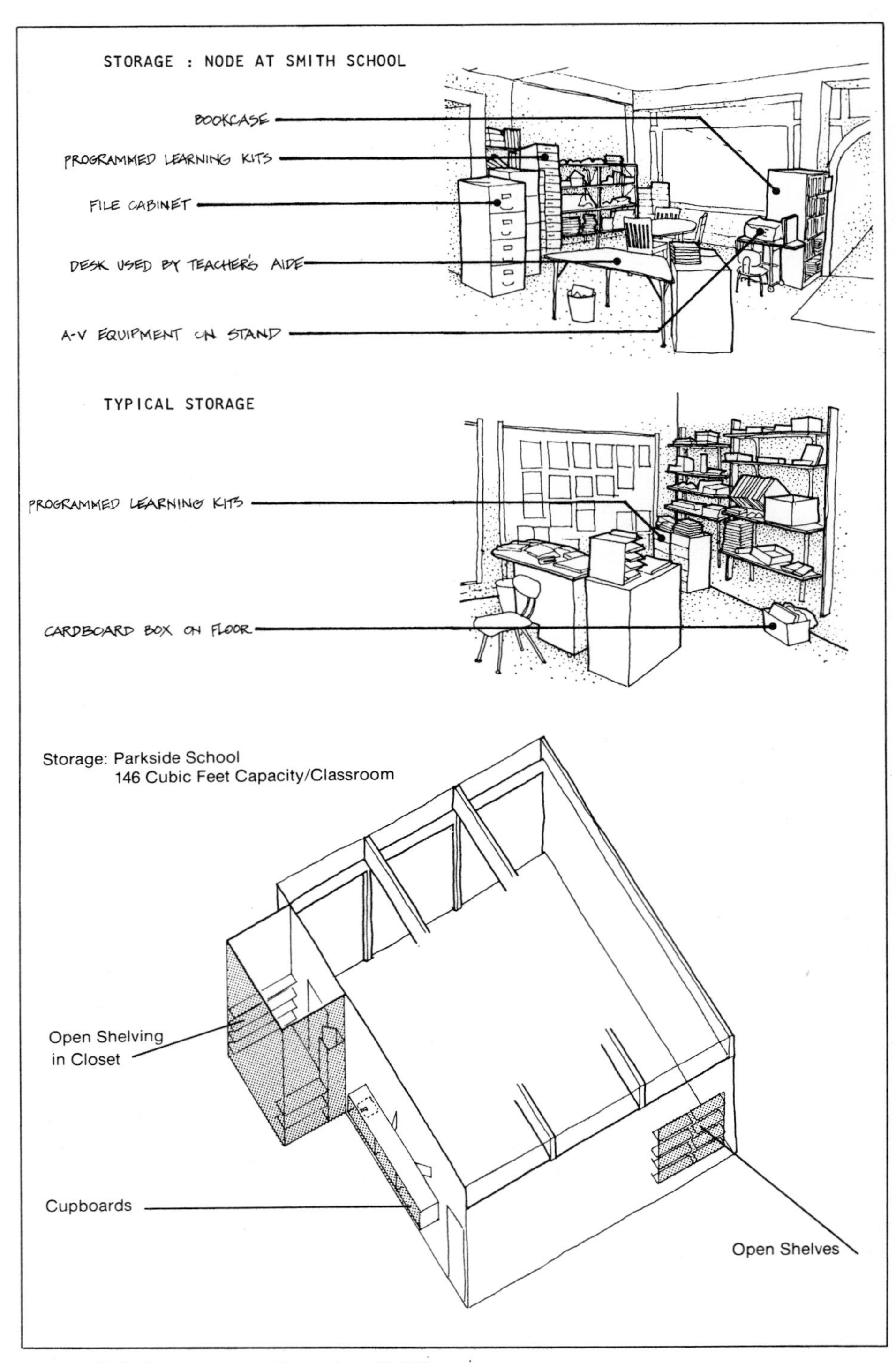

Figure 10-3. Storage capacity and availability.

BEHAVIORAL EVALUATION: TERRITORIALITY

Although there is a growing amount of literature on technical and functional criteria for many building types, there are few criteria that pertain to the behavioral elements of building performance. While schools may have educational objectives and criteria, there are no matching performance criteria on which architectural design can be based. The overall administration and curriculum were common to all of the schools evaluated. However, each school had its own personality, partially due to building design (traditional versus open plan schools), the principal, and the teachers.

The POE investigated relationships or patterns relating the design of the school buildings to the behavior of their occupants. Behavioral observations were originally scheduled over five weeks, but the observation period was extended to eight weeks because of school holidays, school schedules, and missed observations. Four local persons were hired to observe, record, and photograph behavior. For that purpose, standard data-recording sheets and observation schedules were developed (fig. 10-4). The observers were trained in recording data and using unobtrusive automatic cameras. The second-, fourth-, and sixth-grade classrooms and other areas, particularly corridors, were observed. Observation periods were randomly assigned to all of the observers.

The extended observation period helped capture potential changes in activities due to better spring weather conditions. During the on-site practice

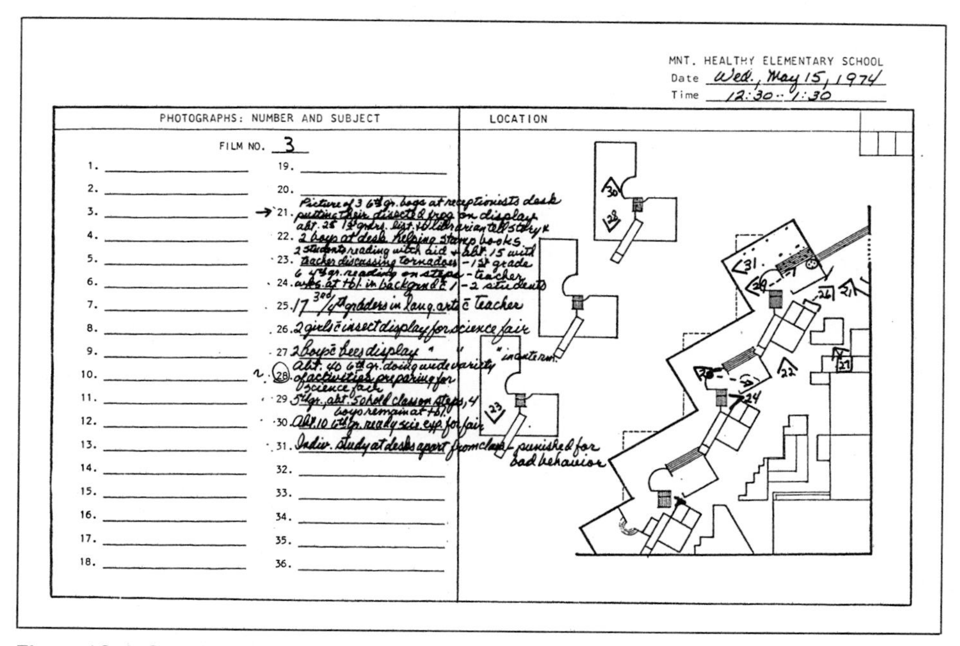
MNT. HEALTHY ELEMENTARY SCHOOL
Date Wed., May 15, 1974
Time 12:30 - 1:30

PHOTOGRAPHS: NUMBER AND SUBJECT | LOCATION

FILM NO. 3

1. – 20. (blank)
21. Picture of 3 6th gr. boys at receptionists desk putting their dissected frog on display
22. 2 boys at desk helping stamp books
23. 2 students reading with aid + abt. 15 with teacher discussing tornadoes - 1st grade
24. 6 4th gr. reading on steps - teacher wks. at tbl. in backgrnd c̄ 1 - 2 students
25. 17 3rd/4th graders in lang. arts c̄ teacher
26. 2 girls c̄ insect display for science fair
27. 2 boys c̄ bees display " " " in anteroom
28. abt. 40 6th gr. doing wide variety of activities preparing for science fair
29. 5th gr., abt. 50 hold class on steps, 4 boys remain at tbl.
30. abt. 10 6th gr. ready scie. exp. for fair
31. Indiv. study at desks apart from class - punished for bad behavior
32. – 36. (blank)

Figure 10-4. Standard data-recording sheet.

runs, some features of the standard observation sheets were changed to emphasize certain behavioral elements of performance and to document activities in some spaces more thoroughly. One change, for instance, concerned the unexpectedly high level of academic activities in some of the corridors, as well as the use of the corridor coat-storage areas for seating and work. Data-collection procedures were modified to monitor these activities more carefully. Photographs and collected data were continuously reviewed throughout the on-site part of the POE process. Examples of results are shown in figure 10-5. Slides of certain activity areas were traced (fig. 10-5), and the observed patterns of behavior commented upon.

Figure 10-5. Illustrations of observed behavior patterns.

3. APPLYING THE POE

3.1. Reporting Findings

This diagnostic POE was both comprehensive and in-depth. To a great extent, it was a research study and included considerable detail. To make the POE comprehensible to a large audience, three levels of detail were presented in the major reports. In addition, an hour-long slide presentation highlighted the major findings of the study.

The first level was the most comprehensive. It included a chart (fig. 10-6) and a few paragraphs summarizing the results in a specific category of the POE, such as acoustics. The chart indicated the performance of the four evaluated schools in terms of ten acoustical areas. It is an oversimplification—there were many types of spaces in each school with varying degrees of performance—but it directed the reader to the most critical issues.

SUMMARY OF PERFORMANCE	P	R	S	M
AMBIENT NOISE				
Unoccupied Classroom	○	○	○	○
Unoccupied Large Spaces	○	○	○	⊙
Occupied Classroom	○	○	○	○
Occupied Large Spaces	⊙	⊙	○	○
Lighting and Mechanical Equipment	⊙	○	○	⊙
TRANSMISSION				
Between Classrooms	⊙	○	○	NA
Hallway/Classrooms	⊙	⊙	⊙	NA
Multipurpose Room/Classroom	NA	⊙	NA	NA
REVERBERATION				
Classroom	○	○	○	○
Large Spaces	●	●	⊙	○

Figure 10-6. Summary of acoustic performance.

A second level included specific findings for elements within any category. For example, acoustical performance tests, including ambient noise, transmission, and reverberation, were conducted and reported in each type of room at each school.

In addition, during a discussion session findings were explained in some detail, and the factors that may have caused the particular performance were elucidated. At times, discussion was lengthy and detailed, depending on the importance and nature of the issue, and graphs and technical details were included (figs. 10-7 and 10-8).

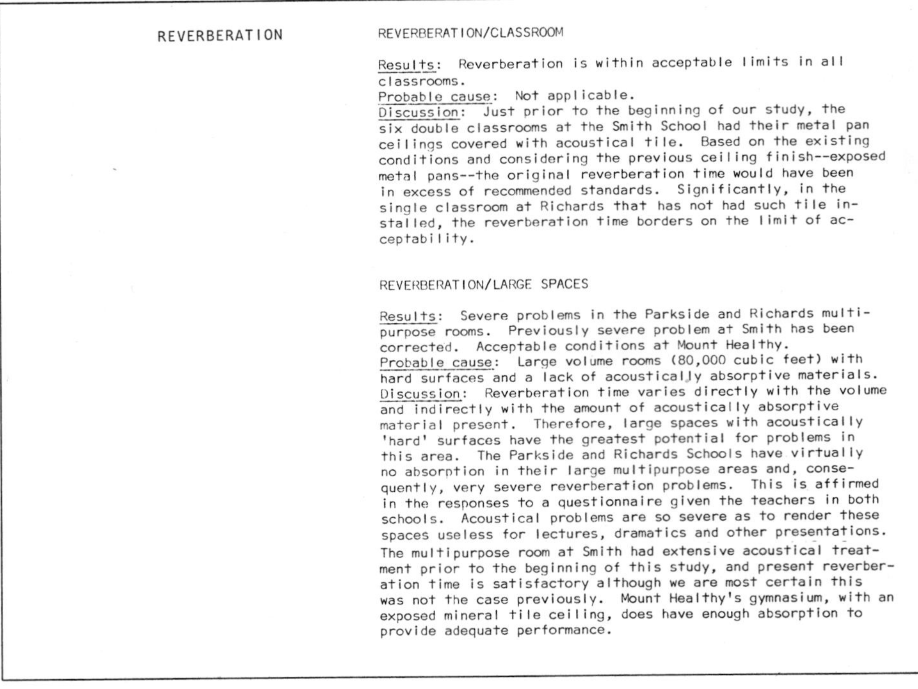

REVERBERATION

REVERBERATION/CLASSROOM

Results: Reverberation is within acceptable limits in all classrooms.
Probable cause: Not applicable.
Discussion: Just prior to the beginning of our study, the six double classrooms at the Smith School had their metal pan ceilings covered with acoustical tile. Based on the existing conditions and considering the previous ceiling finish--exposed metal pans--the original reverberation time would have been in excess of recommended standards. Significantly, in the single classroom at Richards that has not had such tile installed, the reverberation time borders on the limit of acceptability.

REVERBERATION/LARGE SPACES

Results: Severe problems in the Parkside and Richards multipurpose rooms. Previously severe problem at Smith has been corrected. Acceptable conditions at Mount Healthy.
Probable cause: Large volume rooms (80,000 cubic feet) with hard surfaces and a lack of acoustically absorptive materials.
Discussion: Reverberation time varies directly with the volume and indirectly with the amount of acoustically absorptive material present. Therefore, large spaces with acoustically 'hard' surfaces have the greatest potential for problems in this area. The Parkside and Richards Schools have virtually no absorption in their large multipurpose areas and, consequently, very severe reverberation problems. This is affirmed in the responses to a questionnaire given the teachers in both schools. Acoustical problems are so severe as to render these spaces useless for lectures, dramatics and other presentations. The multipurpose room at Smith had extensive acoustical treatment prior to the beginning of this study, and present reverberation time is satisfactory although we are most certain this was not the case previously. Mount Healthy's gymnasium, with an exposed mineral tile ceiling, does have enough absorption to provide adequate performance.

Figure 10-7. Examples of reverberation measurements.

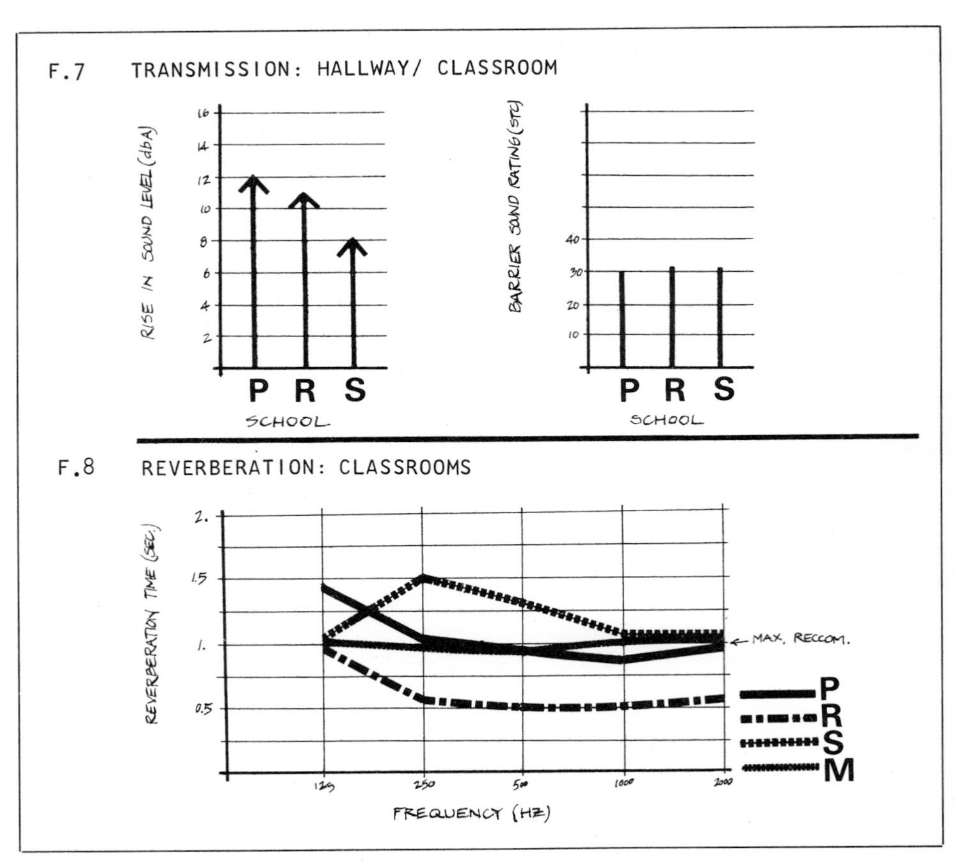

Figure 10-8. Examples of reverberation assessments.

3.2. Recommending Action

Recommendations were not part of the original POE project but were requested by the district school superintendent after the project was completed. The recommendations for action were grouped into three categories, based on the severity of the problems encountered:

- Immediate-term problems that could be resolved by school maintenance personnel
- Medium-term problems that required outside contractors
- Long-term problems that required contractors and consultants and major expenditures

3.3. Reviewing Outcomes

The first group of recommendations discussed above was implemented by the school board's own maintenance and repair personnel with a year of completing the POE.

A surprising result of the evaluation was the wide dissemination of the three-volume POE report (Rabinowitz 1975). Within a year, some two hundred copies of the report had been disseminated through the School of Architecture at the University of Wisconsin–Milwaukee. A mention of the report in an educational facilities journal a year later produced a surge of requests. Since the report was first produced, over one thousand copies have been distributed, and demand is still steady.

ACKNOWLEDGMENTS

The cooperation and enthusiasm of the administration, staff, and students of the Bartholomew County School District, Indiana, made the project work. The help of Francis Krupka, the research assistant on the project, was invaluable during all of the project phases.

Appendices

Appendix A: Legal Issues in Post-Occupancy Evaluation

Robert Greenstreet, Professor
School of Architecture and Urban Planning
University of Wisconsin-Milwaukee

The process of conducting a POE may appear to fall more under the category of applied research and analysis than practice. However, there are several ethical and legal implications to be considered before embarking on such work, and a number of useful strategies and safeguards that can be taken to minimize any potential problem.

The ethical issue of undertaking POEs is intriguing but unclear. Should architects, for example, undertake evaluations of their own buildings, where their objectivity may be impaired or even questioned? More specifically, should architects undertake evaluations that criticize and may even damage their fellow professionals, violating some perceived sense of interfraternal loyalty? The various codes of ethics published by the American Institute of Architects over the years have not really addressed this problem beyond providing the generalized requirement that: "Members should uphold the integrity and dignity of the profession" (*Code of Ethics and Professional Conduct* 1967).

The courts, too, provide little direction (except in the case of inaccurate or malicious reports), although perhaps the analogy of the literary critic is useful in determining the role of the evaluator. The undertaking of a POE, if carried out objectively and skillfully with the degree of performance expected of a professional, should be received in the same light as the work of the theater or book reviewer, who gives a personal, experienced opinion of an actor's or an author's efforts. As such, the evaluator breaks no codes of fidelity of any professional colleagues by the provision of a similar expert service, as long as that service is carried out in a reasonable, professional manner.

However, the legal implications of POEs are more exacting and, as such, require greater attention than those of literary criticism. The degree to which evaluators expose themselves to legal action will depend largely upon the nature of each POE. The provision of a basic factual appraisal of a building, for example, is likely to be less controversial or problematic than a detailed analysis of problems or a diagnosis of failure. In the latter two situations, the evaluator plays a role akin to that of the expert witness, identifying a problem and (explicitly or implicitly) a person or persons responsible. As the evaluator is providing an expert professional opinion in undertaking POEs, it is fair to assume that he or she can be held responsible for the opinion if it is inaccurate or improperly prepared, as in *Hedley Byrne & Co. Ltd.* v. *Heller & Partners Ltd.* 1963–64, United Kingdom. Thus, the employer of the evaluator can expect a reasonable standard of professional care in the preparation of each report. Perhaps more critically, third parties who may be affected by the contents of the report can also expect a legal remedy if its contents, by virtue of inaccuracy, lead to proven loss. Recourse in this case will fall under civil law in the area of tort known as defamation, specifically in the field of libel.

Although accuracy in each evaluation is obviously a prime objective, the complex nature of construction provides many opportunities for potential error. A completed building may be readily accessible for evaluation in many areas, and the identification of defects a relatively simple matter. However, many critical regions will be inaccessible (beneath the ground, under the roof, and so on), and even the most glaring inadequacies may be rendered invisible to the investigator. Similarly, identifying latent defects may prove to be no problem to the experienced evaluator, but latent faults that may be inherent within a building may not become apparent for years, even if they are easily accessible at the time of the inspection (*Dutton* v. *Bognor Regis Urban District Council* 1971, United Kingdom).

To minimize potential misunderstandings, the extent and limitations of each POE should be clearly articulated at the outset of the contractual relationship between the evaluator and the client. It must be made explicitly clear that the evaluation, if positive, does *not* provide an implicit warranty of continued good performance, is accurate only within the available framework of investigation, and is certainly only pertinent to the time period prior to the study. Future concerns, such as changes of use or ownership, variable quality of maintenance, or increased building code requirements cannot be foreseen and may radically alter the future performance of the building.

To ensure a meeting of minds between contracting parties as to the responsibilities of the evaluator and the exact nature of the POE, a written contract that explicitly details the duties and expectations of both parties should be drafted. The nature of the POE (the provision of fact or the determination of fault) should be discussed and accurately transmitted to paper, at which time the evaluator should take the opportunity to include such caveats, exclusion clauses, or waivers as appear to be necessary or desirable. Although these have been occasionally questioned or even dismissed by some courts, they do serve to identify the intentions of both parties prior to contract formulation.

In the event that the POE is being conducted by the architect who originally designed the building (although this is unusual), it may be undertaken under an additional-services clause (American Institute of Architects 1977). Alternatively, if a new contract is being drawn up between an independent evaluator and client, it may be desirable to seek legal assistance to ensure contractual clarity. Again, the nature and extent of each POE will affect the degree of caution and preparation taken in establishing contractual relationships. In all cases, the provision of insurance coverage should be carefully considered to provide adequate protection against future claims.

Appendix B: Usable Tools

INTRODUCTION

Appendix B contains a number of tools intended for use and adaptation by the reader. The forms and checklists are grouped into three sections:

Section a, Instruments for Collecting POE Data, contains checklists, questions, factors, or concepts that commonly occur in POEs. The occupant survey can easily be adapted to any building type by filling in the space categories to be evaluated.

Section b, Formats for Reporting POEs, includes a number of forms that have been found useful in putting together POE reports. Forms are presented in the order in which the POE reports are commonly organized.

Section c presents a Handicapped Accessibility Checklist in graphic format for use in visual inspections of existing buildings or as a reference when checking the plan of building designs.

CHECKLIST OF USEFUL DOCUMENTS FOR P.O.E.

CLIENT-RELATED INFORMATION

(1) Client mission statement, organizational chart, and staffing.
(2) Initial program from building.
(3) As-built floor plans (may require up-dating).
(4) Space assignments and schedules.
(5) Building-related accident reports.
(6) Records of theft, vandalism, and security problems.
(7) Maintenance/repair records.
(8) Energy audits or review comments from heating/cooling plant manager.
(9) Any other feedback concerning the building which may be on record.

BUILDING TYPE-RELATED INFORMATION

(1) Identification of selected recent, similar facilities.
(2) Review of programs and other pertinent information on the building type being evaluated.
(3) Identification and assessment of state-of-the-art literature (e.g., technical manuals and design guides).

We would like to know how well your building performs for all those who occupy it. Successes and failures (if any) are considered insofar as they affect occupant health, safety, efficient functioning, and psychological well-being. Your answers will help improve the design of future, similar buildings.

Below please identify successes and failures in the building by responding to the following broad information categories and by referring to documented evidence or specific building areas wherever possible:

1. Adequacy of Overall Design Concept.
2. Adequacy of Site Design.
3. Adequacy of Health/Safety Provisions.
4. Adequacy of Security Provisions.
5. Attractiveness of Exterior Appearance.
6. Attractiveness of Interior Appearance.
7. Adequacy of Activity Spaces.
8. Adequacy of Spatial Relationships.
9. Adequacy of Circulation Area, e.g., lobby, hallways, stairs, etc.
10. Adequacy of Heating/Cooling and Ventilation.
11. Adequacy of Lighting and Acoustics.
12. Adequacy of Plumbing/Electrical.
13. Adequacy of Surface Materials, e.g., floors, walls, ceilings, etc.
14. Underutilized or Overcrowded Spaces.
15. Other, please specify: (e.g., needed facilities currently lacking).

We wish to conduct a post-occupancy evaluation of your building. The purpose of this evaluation is to assess how well the building perfoms for those who occupy it in terms of health, safety, security, functionality, and psychological comfort. The benefits of a post-occupancy evaluation include: identification of good and bad performance aspects of the building, better building utilization, and feedback on how to improve future, similar buildings.

Please respond only to those questions of the following survey that are applicable to you. Indicate your answers by marking the appropriate blanks with an "X".

1. In an average work week, how many hours do you spend in the following types of spaces (specify):

Space A ______________________
Space B ______________________
Space C ______________________
Space D ______________________
Space E ______________________

HOURS	A	B	C	D	E
0 - 5	( )	( )	( )	( )	( )
6 - 10	( )	( )	( )	( )	( )
11 - 15	( )	( )	( )	( )	( )
16 - 20	( )	( )	( )	( )	( )
21 - 25	( )	( )	( )	( )	( )
26 - 30	( )	( )	( )	( )	( )
31 - 35	( )	( )	( )	( )	( )
35 - 40	( )	( )	( )	( )	( )
40 +	( )	( )	( )	( )	( )

KEY FOR THE FOLLOWING QUALITY RATINGS:

EX = Excellent quality
G = Good quality
F = Fair quality
P = Poor quality

2. Please rate the overall quality of the following areas in the building:

	EX	G	F	P
a) Space Category A	( )	( )	( )	( )
b) Space Category B	( )	( )	( )	( )
c) Space Category C	( )	( )	( )	( )
d) Space Category D	( )	( )	( )	( )
e) Space Category E	( )	( )	( )	( )
f) Restroom(s)	( )	( )	( )	( )
g) Storage	( )	( )	( )	( )
h) Elevator(s)	( )	( )	( )	( )
i) Stairs/Corridors	( )	( )	( )	( )
j) Parking	( )	( )	( )	( )
k) Other, specify________	( )	( )	( )	( )
______________________	( )	( )	( )	( )

3. Please rate the overall quality of Space Category A in terms of the following:

	EX	G	F	P
a) Adequacy of Space	( )	( )	( )	( )
b) Lighting	( )	( )	( )	( )
c) Acoustics	( )	( )	( )	( )
d) Temperature	( )	( )	( )	( )
e) Odor	( )	( )	( )	( )
f) Esthetic Appeal	( )	( )	( )	( )
g) Security	( )	( )	( )	( )
h) Flexibility of Use	( )	( )	( )	( )
i) Other, specify________	( )	( )	( )	( )
______________________	( )	( )	( )	( )

4. Please rate the overall quality of Space Category B in terms of the following:

	EX	G	F	P
a) Adequacy of Space............	( )	( )	( )	( )
b) Lighting............................	( )	( )	( )	( )
c) Acoustics...........................	( )	( )	( )	( )
d) Temperature......................	( )	( )	( )	( )
e) Odor.................................	( )	( )	( )	( )
f) Esthetic Appeal.................	( )	( )	( )	( )
g) Security............................	( )	( )	( )	( )
h) Flexibility of Use.................	( )	( )	( )	( )
i) Other, specify_______........	( )	( )	( )	( )
______________________..	( )	( )	( )	( )

5. Please rate the overall quality of Space Category C in terms of the following:

	EX	G	F	P
a) Adequacy of Space............	( )	( )	( )	( )
b) Lighting............................	( )	( )	( )	( )
c) Acoustics...........................	( )	( )	( )	( )
d) Temperature......................	( )	( )	( )	( )
e) Odor.................................	( )	( )	( )	( )
f) Esthetic Appeal.................	( )	( )	( )	( )
g) Security............................	( )	( )	( )	( )
h) Flexibility of Use.................	( )	( )	( )	( )
i) Other, specify_______........	( )	( )	( )	( )
______________________..	( )	( )	( )	( )

6. Please rate the overall quality of Space Category D in terms of the following:

	EX	G	F	P
a) Adequacy of Space............	( )	( )	( )	( )
b) Lighting............................	( )	( )	( )	( )
c) Acoustics...........................	( )	( )	( )	( )
d) Temperature......................	( )	( )	( )	( )
e) Odor.................................	( )	( )	( )	( )
f) Esthetic Appeal.................	( )	( )	( )	( )
g) Security............................	( )	( )	( )	( )
h) Flexibility of Use.................	( )	( )	( )	( )
i) Other, specify_______........	( )	( )	( )	( )
______________________..	( )	( )	( )	( )

7. Please rate the overall quality of Space Category E in terms of the following:

	EX	G	F	P
a) Adequacy of Space............	( )	( )	( )	( )
b) Lighting............................	( )	( )	( )	( )
c) Acoustics...........................	( )	( )	( )	( )
d) Temperature......................	( )	( )	( )	( )
e) Odor.................................	( )	( )	( )	( )
f) Esthetic Appeal.................	( )	( )	( )	( )
g) Security............................	( )	( )	( )	( )
h) Flexibility of Use.................	( )	( )	( )	( )
i) Other, specify_______........	( )	( )	( )	( )
______________________..	( )	( )	( )	( )

8. Please rate the overall quality of design in this building:

	EX	G	F	P
a) Esthetic quality of exterior........	( )	( )	( )	( )
b) Esthetic quality of interior.........	( )	( )	( )	( )
c) Amount of space......................	( )	( )	( )	( )
d) Environmental quality (lighting, acoustics, temperature, etc.)......	( )	( )	( )	( )
e) Proximity to views......................	( )	( )	( )	( )
f) Adaptability to changing uses.....	( )	( )	( )	( )
g) Security....................................	( )	( )	( )	( )
h) Maintenance............................	( )	( )	( )	( )
i) Relationship of spaces/layout.....	( )	( )	( )	( )
j) Quality of building materials........				
(1) Floors..............................	( )	( )	( )	( )
(2) Walls...............................	( )	( )	( )	( )
(3) Ceilings...........................	( )	( )	( )	( )
k) Other, specify...____________	( )	( )	( )	( )
____________________________	( )	( )	( )	( )

9. Please select and rank in order of importance facilities which are currently lacking in your building:

10. Please make any other suggestion you wish for physical or managerial improvements in your building:

11. Demographic Information:

a) Your Room #/Building area ____________________

b) Your Position: ____________________

c) Your Age: ________

d) Your Sex: ________

e) # of years with the present organization: _______

PERFORMANCE CRITERIA MET:	Yes √	No √
STABILITY	( )	( )
MOVEMENT	( )	( )
• STRUCTURAL LOADING	( )	( )
• THERMAL MOVEMENT	( )	( )
• SETTING	( )	( )
IMPACT	( )	( )
AIR INFILTRATION	( )	( )
MOISTURE INFILTRATION	( )	( )
THERMAL CONDUCTIVITY	( )	( )
STAINING	( )	( )
DISCOLORATION	( )	( )
DELAMINATION	( )	( )
DETERIORATION	( )	( )
ESTHETICS	( )	( )

PERFORMANCE CRITERIA MET:	Yes √	No √
DRAINAGE (ponding)	( )	( )
MOISTURE PENETRATION	( )	( )
SAG	( )	( )
MOVEMENT	( )	( )
DETERIORATION	( )	( )
EROSION	( )	( )
IMPACT	( )	( )
INDENTATION	( )	( )
BRITTLENESS	( )	( )

PERFORMANCE CRITERIA MET:	Yes √	No √
STRUCTURAL STABILITY	( )	( )
IMPACT	( )	( )
ATTACHED LOADS	( )	( )
COHESION	( )	( )
DELAMINATION	( )	( )
WEARABILITY	( )	( )
INDENTATION	( )	( )
ABRASION	( )	( )
SCRATCH	( )	( )
WATER ABSORPTION	( )	( )
STAIN	( )	( )
CLEANABILITY	( )	( )
REPLACEMENT/REPAIR	( )	( )
ESTHETICS	( )	( )

PERFORMANCE CRITERIA MET:	Yes √	No √
DEFLECTION	( )	( )
DISPLACEMENT	( )	( )
COHESION	( )	( )
INDENTATION (impact)	( )	( )
SCRATCH	( )	( )
STAINING	( )	( )
COLOR HOMOGENEITY	( )	( )
FLAKING/PEELING	( )	( )
FADING	( )	( )
CLEANABILITY	( )	( )
ACCESS TO PLENUM	( )	( )
REPLACEMENT/REPAIR	( )	( )

PERFORMANCE CRITERIA MET:	Yes √	No √
INDENTATION (impact)	( )	( )
SCRATCH	( )	( )
STAINING	( )	( )
COLOR HOMOGENEITY	( )	( )
RESILIENCY	( )	( )
COLOR FASTNESS/FADING	( )	( )
CLEANABILITY	( )	( )
CIGARETTE BURN	( )	( )
REPLACEMENT/REPAIR	( )	( )
WATER ABSORPTION	( )	( )
STATIC DISCHARGE	( )	( )
WEAR	( )	( )
SLIP RESISTANCE	( )	( )
ADHESION	( )	( )
LEVELNESS	( )	( )

TECHNICAL FACTORS: LIGHTING

USABLE TOOLS

PERFORMANCE CRITERIA MET:	Yes √	No √
ILLUMINATION - NATURAL (f.c.)	( )	( )
ILLUMINATION - ARTIFICIAL	( )	( )
ILLUMINATION - COMBINED	( )	( )
ILLUMINATION - shades fully drawn	( )	( )
LUMINAIRE LUMINANCE (f.c.)	( )	( )
ROOM CONTRAST RATIO	( )	( )
TASK SURROUND CONTRAST RATIO	( )	( )
LUMINANCE GAIN (cleaning)	( )	( )
DIRECT GLARE	( )	( )

PERFORMANCE CRITERIA MET:	Yes √	No √
AMBIENT SOUND LEVEL (d.b.) (with users present)	( )	( )
AMBIENT SOUND LEVEL (d.b.) (without users present)	( )	( )
AMBIENT SOUND LEVEL (d.b.) (without users, lights)	( )	( )
ATTENUATION (d.b.)	( )	( )
REVERBERATION TIME 500, 1000, 2000 hz	( )	( )
MECHANICAL SYSTEMS	( )	( )
IMPACT GENERATED NOISE	( )	( )

USABLE TOOLS

PERFORMANCE CRITERIA MET:	Yes √	No √
AMBIENT TEMPERATURE	( )	( )
TEMPERATURE GRADIENT	( )	( )
HUMIDITY	( )	( )
AIR MOVEMENT	( )	( )
SAFETY HAZARDS	( )	( )

01 Access
02 Adaptation
03 Ambiguity
04 Animation
05 Attitude
06 Boredom
07 Communication
08 Comfort
09 Congruence
10 Competence
11 Complexity
12 Continuity
13 Control
14 Crowding
15 Culture
16 Density
17 Distance
18 Diversity
19 Dominance
20 Exposure
21 Familiarity
22 Habitability
23 Identification
24 Image
25 Integrity
26 Interaction
27 Interpersonal Relations
28 Isolation
29 Manipulation
30 Mobility
31 Novelty
32 Order
33 Orientation (direction finding)
34 Personalization
35 Privacy
36 Projection
37 Proximity
38 Security
39 Sensory Deprivation
40 Sensory Stimulation
41 Status
42 Stress
43 Territoriality
44 Unity
45 Variability, Variety

POE PROJECT OVERVIEW

1. Level of Effort:
2. Purpose/Use:
3. Building Type:
4. Building Site:
5. Building Age:
6. Building Location(s):
7. Building Client:
8. Important Factors Analyzed:
   - Technical
   - Functional
   - Behavioral
9. Methods Used:
   - Data Collection
   - Data Analysis
10. Project Personnel:
11. Project Duration/Manhours:
12. Project Cost:
13. Major Lessons Learned:
14. Benefits to Client:

HOW TO READ THIS BOOK

◊ OPEN THE BOOK COMPLETELY.
◊ SPECIFIC BUILDING AREA IS INDICATED AT UPPER LEFT.

◊ EVALUATION STATEMENTS REFER TO SPECIFIC AREAS DESCRIBED ON FACING PAGE.

◊ SPECIFIC AREA REQUIREMENTS ARE LISTED BELOW (SEE GLOSSARY FOR DEFINITION OF TERMS).

◊ EVALUATION STATEMENTS TYPICALLY REFER TO ITEMS WHICH ARE DETERMINED TO BE SUBSTANDARD AND ARE PRESENTED TOGETHER WITH RECOMMENDATIONS.

CRITERIA

1.0 OCCUPANTS

2.0 ACTIVITIES/TIME

3.0 OCCUPANT REQUIREMENTS
3.1 HEALTH/SAFETY & SECURITY REQUIREMENTS

3.2 FUNCTIONAL REQUIREMENTS

3.3 PSYCHOLOGICAL REQUIREMENTS

4.0 AMBIENT ENVIRONMENT
4.1 HEAT/COOL:
4.2 OLFACTORY:
4.3 VENTILATION:
4.4 LIGHTING:
4.5 ACOUSTICS:

5.0 LOCATIONAL REQUIREMENTS

6.0 OCCUPANT-EQUIPMENT REQUIREMENTS

7.0 SPECIAL REQUIREMENTS

8.0 MATERIALS/FINISHES

9.0 CODES/AGENCY REQUIREMENTS

10. AREA

EVALUATION

◊ SELECTED CRITERIA (SEE FACING PAGE) REQUIREMENT NUMBERS RECUR IN EVALUATION FOR EASY CROSS-REFERENCING.

9. ETC.

B. FORMATS FOR REPORTING POES

EXECUTIVE SUMMARY

INTRODUCTION

METHODOLOGY

CRITERIA

1.0 OCCUPANTS

2.0 ACTIVITIES/TIME

3.0 OCCUPANT REQUIREMENTS

3.1 HEALTH/SAFETY & SECURITY REQUIREMENTS

3.2 FUNCTIONAL REQUIREMENTS

3.3 PSYCHOLOGICAL REQUIREMENTS

4.0 AMBIENT ENVIRONMENT

4.1 HEAT/COOL:

4.2 OLFACTORY:

4.3 VENTILATION:

4.4 LIGHTING:

4.5 ACOUSTICS:

5.0 LOCATIONAL REQUIREMENTS

6.0 OCCUPANT-EQUIPMENT REQUIREMENTS

7.0 SPECIAL REQUIREMENTS

8.0 MATERIALS/FINISHES

9.0 CODES/AGENCY REQUIREMENTS

10. AREA

CRITERIA

EVALUATION

EVALUATION

STATEMENT:

RESPONSE

STRONGLY AGREE										
AGREE										
NOT SURE										
DISAGREE										
STRONGLY DISAGREE										
	10	20	30	40	50	60	70	80	90	100

PERCENT

MEAN:

STANDARD DEVIATION:

STATEMENT:

RESPONSE

STRONGLY AGREE										
AGREE										
NOT SURE										
DISAGREE										
STRONGLY DISAGREE										
	10	20	30	40	50	60	70	80	90	100

PERCENT

MEAN:

STANDARD DEVIATION:

APPENDIX

C. HANDICAPPED ACCESSIBILITY CHECKLIST

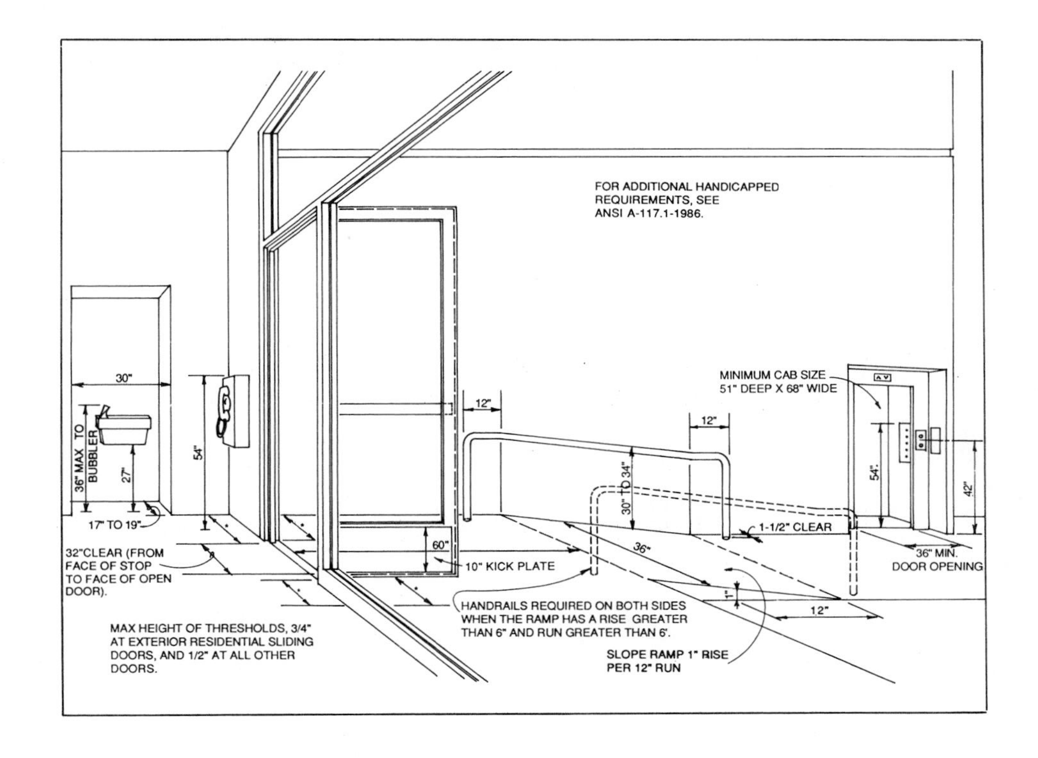

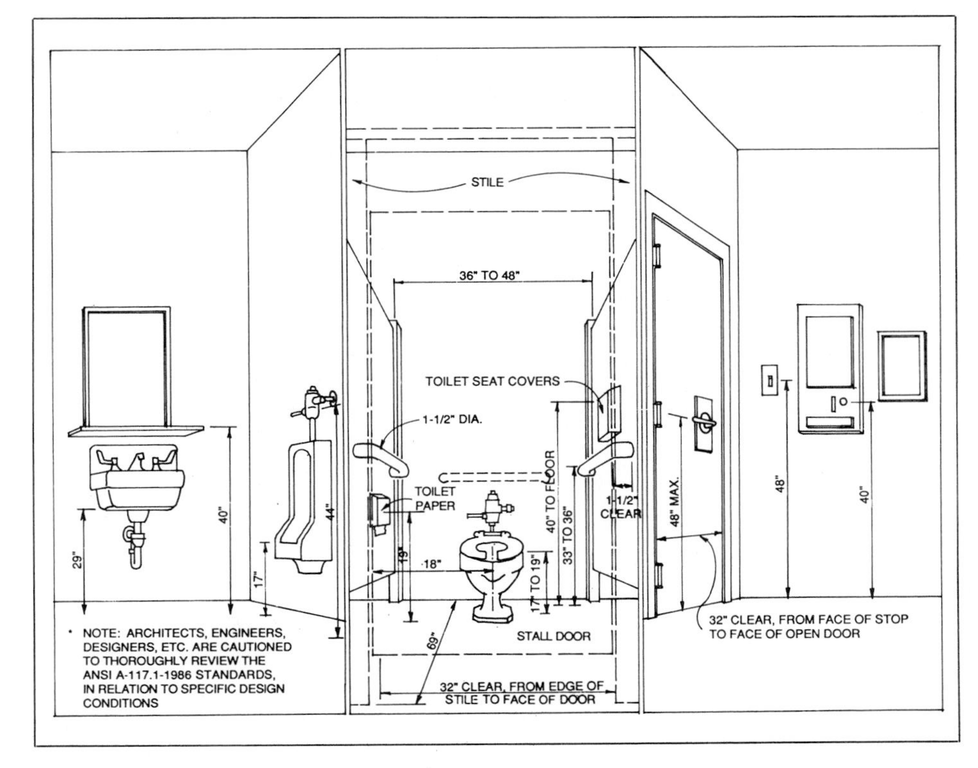

Appendix C: Resources

MAJOR ORGANIZATIONS WITH POE ACTIVITY

American Society for Testing and Materials (ASTM)
1916 Race Street
Philadelphia, Pennsylvania 19103

Architecture and Engineering Performance Information Center (AEPIC)
3907 Metzerot Road
University of Maryland
College Park, Maryland 20742

Building Officials Management Association (BOMA)
1250 I Street, N.W.
Suite 200
Washington, D.C. 20005

Environmental Design Research Association (EDRA)
Care of Willo P. White, Ph. D.
L'Enfant Plaza Station
P.O. Box 23129
Washington, D.C. 20026

International Association for the Study of People and Their Physical Surroundings (IAPS)
Care of Dr. Martin Symes
Bartlett School of Architecture and Planning
University College London
London WC1E 6BT, England

International Facilities Management Association (IFMA)
Summit Tower, Suite 1410
11 Greenway Plaza
Houston, Texas 77046

Man-Environment Research Association (MERA)
Care of Professor Ichiro Souma
Department of Psychology, Faculty of Literature
Waseda University
1-24-1 Toyama, Shinjuku-Ku
Tokyo 162, Japan

People and the Physical Environment Research Association (PAPER)
Care of Professor Ross Thorne
Architectural Psychology Research Unit, Faculty of Architecture
University of Sydney
Sydney NSW 2006, Australia

MAJOR DATABASES

American Institute of Architects Research Information Retrieval System
1735 New York Avenue, N.W.
Washington, D.C. 20006
Referral service for architects seeking access to researchers working on specific problems.

Architecture and Engineering Performance Information Center (AEPIC)
3907 Metzerot Road
University of Maryland
College Park, Maryland 20742

ARKISYST
International Union of Architects
1, rue d'Ulm
75005 Paris
France
A worldwide effort sponsored by the International Union of Architects to assess the total information environment of the architectural community.

ASMER, Inc.
Care of Aristide H. Esser, M.D.
P.O. Box 57
Orangeburg, New York 10962
A feasibility study on improving access to information about the impact of the built environment on human life and health.

DATRIX
Xerox University Microfilms
300 North Zeeb Road
Ann Arbor, Michigan 48106
Dissertation abstracts file of Ph.D. dissertations in all subject areas.

Educational Resources Information Center (ERIC)
National Institute of Education
300 - 7th Street, S.W.
Washington, D.C. 20202
Journal articles and reports, including ongoing research in the field of education. For example, ERIC Clearinghouse in Educational Management covers all aspects of educational facilities.

Human Engineering Information Analyses Center
U.S. Army Aberdeen Research & Development Center
Aberdeen Proving Ground, Maryland 21005
All aspects of human factors research. Access limited to federal agencies and government contractors.

Institute for Scientific Information
325 Chestnut Street
Philadelphia, Pennsylvania 19106
Publishes "Current Contents—Behavioral Social & Education Services" and "Social Services Citation Index." Offers computerized searches of these files.

National Referral Center
Library of Congress
10 First Street, S.E.
Washington, D.C. 20540
An "information desk" directs user to organizations or individuals with specialized knowledge on a subject.

(continued)

National Technical Information Service
U. S. Department of Commerce
5281 Port Royal Road
Springfield, Virginia 22151
File covers U.S. government-sponsored research reports. Publishes "Government Reports Announcements" and "Government Reports Index."

Psychological Abstracts
American Psychological Association
1200 17th Street, N.W.
Washington, D.C. 20036
Journal literature and books on all aspects of psychology. Publishes *Psychological Abstracts*.

Smithsonian Science Information Exchange
1730 M Street, N.W.
Washington, D.C. 20036
Current information on research in progress in behavioral and physical sciences.

Vance Bibliographies
P.O. Box 229
112 North Charter Street
Monticello, Illinois 61856
A series of topically focused bibliographies in architecture and planning.

Glossary

Access. The route taken from a given location (street, parking lot, and the like) to the site or the building entry.
Activities. Those events or behaviors taking place within a given area by a specific type of occupant.
Adjacency. The spatial relationships of adjoining spaces or areas to each other. Implied are common building surfaces, shared access, and other physical aspects of adjoining spaces.
Ambient environment. The sensory properties of a given area. Vision, hearing, sense of smell, and temperature are considered, among others.
Chi-square. Statistics for the analysis of differences between two groups when the data are normal—that is, categorically different (for example, Republicans and Democrats).
Circulation. The physical process of moving persons or goods through a building. Circulation areas refer to corridors, elevators, ramps, stairs, or any other undesignated activity areas.
Communication format. The narrative, graphic means, or media for conveying information to client groups or organizations.
Criteria. The standards against which the performance of actual buildings in use is compared.
Cybernetics. The study of human control functions and of mechanical and electrical systems designed to replace them.
Data collection or gathering. The process by which information pertinent to a given study is obtained. Data are extracted from a variety of sources.
Decay time. The time it takes for a sound to diminish a specific amount.
Documents. All printed matter examined in a study, including institutional publications, accident reports, newspaper accounts, and so on.
Environment. Denotes the physical attributes of a given activity area.
Evaluation. The activity of comparing performance criteria with actual performance measures.
Factor analysis. A multivariate method used to help the researcher discover and identify the effects of each of a number of variables, called factors.
Formal. The image and symbolic characteristics of a building. For example, the physical design articulation of an entry would be a formal element.
Function. Denotes specific area/activity requirements for a given setting. For example, an office setting might have the functional requirements of photocopying space, desk space, restrooms, and waiting area.

Generic setting. The physical setting category for specific types of activities, such as sleeping in "bedroom," hygiene in "bathroom," and so on.

Habitability. The quality of a space in terms of occupant needs such as health, safety, security, functional effectiveness, and psychological comfort and satisfaction.

Handicap. Any social, emotional, or physical disability. As an adjective, as in "handicapped access," it usually refers to specific criteria to facilitate use by an individual who has a disability.

Image. The perception of a given building or environment in terms of esthetic quality.

Inter-observer reliability. The carefully orchestrated comparisons of observations made by different observers of the same scene, that is, a check of whether the same phenomena are being observed.

Intuitive evaluation. That aspect of an evaluation not substantiated by objective data, but based on the intuitive judgment and expertise of the evaluator.

Location. The physical placement of a specific area.

Mean. The accumulated average of the sum.

Methodological critique. The review of particular strengths and weaknesses of the evaluation process.

Observations. The phenomena physically observed on site by the evaluators.

Occupant. Any person using a given building or area.

Orientation. The process of determining one's location on a site or in a building.

Performance. The ability of an environment to support occupant requirements as described by evaluation criteria.

Physical. Construction elements in the built environment.

Post-construction evaluation. An evaluation primarily concerned with the physical performance of a building after completion of construction.

Post-occupancy evaluation. The process of systematic data collection, analysis, and comparison with explicitly stated peformance criteria pertaining to occupied, built environments.

Proximity. The physical positioning of areas close to each other.

Psychological. The emotional and intellectual effect of a building or area as perceived by its occupants.

Questionnaire. The written survey used to gather subjective information from occupants regarding building performance.

Regression analysis. A statistical technique that attemps to predict the value of a variable from its relationship with another variable.

Reverberation testing. A means of assessing the acoustical properties of a room by measuring the length of time that a sound persists because it is being deflected off of the room's surfaces.

Satisfaction. Acceptable accommodation of occupant/user needs.

Stepwise regression. A statistical technique that pinpoints closely related variables when many variables are used.

T-test. A measure of the differences between the means of two groups.

Variance. A measure of the variation of the observations within a given population.

Bibliography

Ahrentzen, S., et al. "School Environment and Stress." In *Environmental Stress,* edited by G. W. Evans. New York: Cambridge University Press, 1982.

Alexander, C. *Notes on the Synthesis of Form.* Cambridge, MA: Harvard University Press, 1964.

Alexander, C., et al. *Houses Generated by Patterns.* Berkeley: Center for Environmental Structure, 1969.

———. *A Pattern Language: Towns—Buildings—Construction.* New York: Oxford University Press, 1977.

American Institute of Architects. Article III, *Ethical Principles,* AIA document 6-J-400, Washington, DC, March 10, 1981.

———. *Owner-Architect Agreement,* AIA document B-141, 13th edition, Article 1.7.22., Washington DC, 1977.

American Society for Testing and Materials (ASTM). *Standard Practice for Rating the Overall Performance of an Existing Building or Facility.* Draft Document, ASTM Subcommittee E6.25 (Gerald Davis, Chairman). Philadelphia: May 1986.

"Basic Bedrooms: How Marriott Changes Hotel Design to Tap Mid-Priced Market." *Wall Street Journal.* 18 September 1985.

Bechtel, R. *What Are Post-Occupancy Evaluations? A Layman's Guide to POE for Housing.* Final Draft Report. Tucson, AZ: Environmental Research and Development Foundation, 1980.

Becker, F. D. *Design for Living—The Resident's View of Multi-Family Housing.* Ithaca, NY: Center for Urban Development Research, Cornell University, 1974.

Bonshor, R. B., and H. W. Harrison. *An Investigation into Faults and Their Avoidance.* Quality in Traditional Housing, vol. 1. London: Department of the Environment, Building Research Establishment, HMSO, 1982.

Brill, M., et al. *Using Office Design to Increase Productivity,* vol. 1. Buffalo, NY: Workplace Design and Productivity Inc., 1984.

Building Research Board. *Building Diagnostics: A Conceptual Framework.* Washington, DC: National Academy Press, 1983.

———. *Programming Practices in the Building Process—Opportunities for Improvement.* Washington, DC: National Academy Press, 1986.

———. *Post-Occupancy Evaluation in the Building Process—Opportunities for Improvement.* Washington, DC: National Academy Press, 1987.

Canter, D. *Architectural Psychology: Proceedings of the Conference Held at Dalandhui.* London: RIBA Publications Ltd., 1970.

Clipson, C., and J. Wehrer. *Planning for Cardiac Care.* Ann Arbor, MI: The Health Administration Press, University of Michigan, 1973.

Cooper, C. *Residents' Attitudes Toward the Environment at St. Francis Square, San Francisco.* Berkeley: University of California, Institute of Urban and Regional Development, 1970.

———. *Easter Hill Village: Some Social Aspects of Design.* New York: The Free Press, 1975.

Craighead, J. E., and Mossman, B. T. "The Pathogenesis of Asbestos-Associated Diseases." *New England Journal of Medicine* 306 (1982).

Daish, J., J. Gray, and D. Kernohan. *Post-Occupancy Evaluation of Government Buildings.* Wellington, New Zealand: Victoria University of Wellington, School of Architecture, 1980.

———. *Post-Occupancy Evaluation Trial Studies,* nos. 1–3. Wellington, New Zealand: Victoria University of Wellington, School of Architecture, 1981.

Davis, T. A. "Evaluating for Environmental Measures." In *Proceedings of the Second Annual Environmental Design Research Association Conference,* edited by J. Archea and C. Eastman. Pittsburgh: Carnegie-Mellon University, 1970.

Department of the Army. *Design Guide DG 1110–3–106 US Army Service Schools.* Washington, DC: Engineering Division, Military Construction Directorate, Office of the Chief of Engineers, 1976.

Diekmann, J. E., and M. C. Nelson. "Construction Claims: Frequency and Severity." *The Journal of Construction, Engineering, and Management* 3, no. 1 (March 1985).

Dutton v. *Bognor Regis Urban District Council* (1971) 2 All ER 103.

Eberhard, J. P. "Horizons for the Performance Concept in Building." in Building Research Board, *Proceedings of the Symposium on the Performance Concept in Building.* Washington, DC: National Academy of Sciences, 1965.

Ehrenkrantz, Ezra. *SCSD: The Project and the Schools.* New York: Educational Facilities Laboratories, Inc., 1967.

Eshelman, M. P., W. F. E. Preiser, et al. *Albuquerque High School—A Post-Occupancy Evaluation.* Institute for Environmental Education Monograph Series no. 9. Albuquerque: University of New Mexico, 1981.

Field, H., et al. *Evaluation of Hospital Design: A Holistic Approach.* Boston: Tufts–New England Medical Center, 1971.

Francescato, G., S. Weidemann, and J. R. Anderson. "Residential Satisfaction: Its Uses and Limitations in Housing Research." In *Housing in an Era of Fiscal Austerity,* edited by W. van Vliet. Westport, CT: Greenwood Press, in press.

Francescato, G., et al. *Residents' Satisfaction in HUD-Assisted Housing: Design and Management Factors.* Washington, DC: U.S. Department of Housing and Urban Development, U.S. Government Printing Office, March 1979.

Friedman, A., C. Zimring, and E. Zube. *Environmental Design Evaluation.* New York: Plenum Press, 1978.

General Services Administration. *The PBS Building Systems Program and Performance Specifications for Office Buildings,* 3rd ed. Washington, DC: Government Printing Office, November 1975.

Goodrich, R. *Post-Design Evaluation of Centre Square Project.* Philadelphia: Atlantic Richfield, Inc., 1976.

Greenstreet, R., and K. Greenstreet. *The Architect's Guide to Law and Practice.* New York: Van Nostrand Reinhold, 1984.

Hall, E. T. *The Hidden Dimension,* Garden City, NY: Doubleday, 1966.

Hall, E. T., and M. Hall. *The Fourth Dimension in Architecture: The Impact of Building on Man's Behavior.* Santa Fe, NM: The Sunstone Press, 1975.

Hedley Byrne & Co. Ltd. v. *Heller & Partners, Ltd.* (1963–64) 2 All ER 575.

Hsia, V. *Residence Hall Environments: An Architectural Psychology Case Study.* Salt Lake City, UT: University of Utah, 1967.

Jordan, J. J. *Senior Center Facilities: An Architect's Evaluation of Building Design, Equipment and Furnishings.* Washington, DC: National Council on Aging, November 1975.

———. *Senior Center Design: An Architect's Discussion of Facility Planning.* Washington, DC: National Council on Aging, March 1978.

Kantrowitz, M., et al. "P/A POE: Energy Past and Future." *Progressive Architecture* (April 1986).

Kirkpatrick, K., et al. *Testing the Program—Bear Canyon Senior Center.* Albuquerque, NM: Center for Research and Development Monograph Series no. 27. School of Architecture and Planning, University of New Mexico, 1986.

Kish, L. *Survey Sampling.* New York: Wiley, 1965.

Leslie, H. G. *Concepts of Project (Building) Evaluation—An Overview.* Canberra, Australia: National Committee on Rationalized Building, 1985.

Manning, P. *Office Design: A Study of Environment.* Liverpool, England: Pilkington Research Unit, Department of Building and Science, University of Liverpool, 1965.

Marans, R., and K. Spreckelmeyer. *Evaluating Built Environments: A Behavioral Approach.* Ann Arbor, MI: The University of Michigan, Institute for Social Research and College of Architecture and Urban Planning, 1981.

Markus, T., et al. *Building Performance.* New York: Halstead Press, 1972.

McLaughlin, H., J. Kibre, and R. Mort. "Patterns of Physical Change in Six Existing Hospitals." In *Environmental Design: Research and Practice,* edited by W. Mitchell. Los Angeles: University of California, 1972.

Nasar, J. L. (ed.). *The Visual Quality of the Environment: Theory, Research, and Application.* Cambridge, England: Cambridge University Press, 1988.

Newman, O. *Defensible Space: Crime Prevention Through Urban Design.* New York: Collier Books, 1973.

Nie, N. H., et al. *Statistical Package for the Social Sciences.* New York: McGraw-Hill, 1970.

Office of Protection from Research Risks. *Protection of Human Subjects.* Code of Federal Regulations 45CFR46–March 8, 1983. Washington, DC: National Institute of Health, 1983.

Oppenheim, N. N. *Questionnaire Design and Attitude Measurement.* New York: Basic Books, 1966.

Orbit-Information Technology and Office Design. London: Duffy, Eley, Giffone, Worthington, Architects and Space Planners, 1983.

Osmond, H. "Some Psychiatric Aspects of Design." In *Who Designs America?,* edited by L. B. Holland. Garden City, NY: Doubleday, 1966.

Parks, G. M. *Economies of Carpeting and Resilient Flooring.* Philadelphia: University of Pennsylvania Press, 1966.

Parshall, S. A., and W. M. Peña. *Evaluating Facilities: A Practical Approach to Post-Occupancy Evaluation.* Houston: CRS Sirrine, Inc., 1982.

Peña, W. M., W. Caudill, and J. Focke. *Problem Seeking: An Architectural Programming Primer.* Boston: Cahners Books, 1977.

Prak, N. L. *Architects: The Noted and the Ignored.* New York: Wiley, 1984.

Preiser, W. F. E. *Behavioral Design Criteria in Student Housing—The Measurement of Verbalized Response to Physical Environment.* Blacksburg, VA: Virginia Polytechnic Institute and State University, 1969.

———. *Behavior and Design: A Core Bibliography,* rev. ed. Architecture Series Bibliography, no. A-24. Monticello, IL: Vance Bibliographies, October 1978.

———. "A Prototype Post-Occupancy Evaluation of the Agricultural Sciences Building–South at the University of Kentucky." In *Proceedings of the Conference on People and Physical Environment Research.* Wellington, New Zealand, June 1983.

———. "The Habitability Framework: Environmental Design Cybernetics as a Conceptual Basis for Person-Environment Relationships." In *Human Ecology: A Gathering of Perspectives.* Washington, D.C.: Society for Human Ecology, 1986.

Preiser, W. F. E., and J. Daish. *Post-Occupancy Evaluation—A Selected Bibliography,* no. A-896. Monticello, IL: Vance Bibliographies, January, 1983.

Preiser, W. F. E., and R. R. Pugh. *Senior Centers.* Albuquerque, NM: Center for Research and Development Monograph Series no. 26, School of Architecture and Planning, University of New Mexico, 1986a.

———. "Senior Centers: A Process Description of Literature Evaluation, Walk-through Post-occupancy Evaluations, A Generic Program and Design for the City of Albuquerque." In *The Costs of Not Knowing: Proceedings of the 17th Annual Environmental Research Design Association Conference,* edited by J. Wineman, R. Barnes, and C. Zimring. Washington, DC: EDRA, Inc., 1986b.

Public Works Canada. *Project Delivery System, Stage 10: Level 1 Evaluation, Users Manual.* Ottawa, Canada: Department of Planning and Coordination Branch, June 1979.

Rabinowitz, H. Z. *Buildings in Use Study.* Milwaukee, WI: University of Wisconsin, School of Architecture and Urban Planning, 1975.

Roethlisberger, F. J., and W. J. Dickson. *Management and the Worker.* Cambridge, MA.: Harvard University Press, 1939.

Sanoff, H. *Techniques of Evaluation for Designers.* Raleigh, NC: North Carolina State University, Design Research Laboratory, 1968.

Scott, G. *Building Disasters and Failures.* Hornby, England: Construction Press, Ltd., 1976.

Slavin, M. "The New Art of Positive Criticism." *Interiors* (April 1982).

Sommer, R. *Personal Space: The Behavioral Basis of Design.* Englewood Cliffs, NJ: Prentice-Hall, 1969.

———. *Tight Spaces: Hard Architecture and How to Humanize It.* Englewood Cliffs, NJ: Prentice-Hall, 1974.

Survey Research Center—Institute for Social Research. *Interviewer's Manual,* rev. ed. Ann Arbor, MI: University of Michigan, 1976.

Thorne, R. *Building Appraisal.* (Videotape, scripted from material supplied by the Building Performance Research Unit, University of Strathclyde, Glasgow.) Sydney, Australia: University of Sydney, Department of Architecture, 1980.

Trites, D. K., et al. "The Influence of Nursing Unit Design on the Activities and Subjective Feelings of Nursing Personnel." *Environment and Behavior* 2, no. 3 (1970).

Van der Ryn, S., and M. Silverstein. *Dorms at Berkeley.* Berkeley: University of California, Center for Planning and Research, 1967.

Webb, E. J., et al. *Unobtrusive Measures: Nonreactive Research in the Social Sciences.* Chicago: Rand McNally, 1966.

Weidemann, S., et al. "Residents' Perception of Satisfaction and Safety: A Basis for Change in Multifamily Housing." *Environment and Behavior* 14, no. 6 (1982).

Weinstein, C. S. "The Physical Environment of the School: A Review of the Research." *Review of Education Research* 49, no. 4 (1979).

White, E. T. *The Value of Post-Occupancy Evaluation to the Architect in Government.* Tallahassee, Florida: School of Architecture, Florida A & M University, 1983.

———. *Building Evaluation in Professional Practice.* Tallahassee: School of Architecture, Florida A & M University, 1985.

Wright, J. R. "Performance Criteria in Building." *Scientific American* 224, no. 3 (March 1971).

Zeisel, J. *Inquiry by Design: Tools for Environment-Behavior Research.* Monterey, CA: Brooks-Cole, 1981.

Index

Page numbers in *italics* refer to figures; page numbers in **boldface** refer to tables.